College Algebra

Student Workbook

Sixth Edition

Dr. Marilyn P. Carlson
Professor, Mathematics Education
Arizona State University

Phoenix

Pathways College Algebra – Student Workbook, Sixth Edition

Published by Rational Reasoning, LLC., PO Box 310, Lansing, KS 66043.

This book was typeset in 11/12 Times Roman

ISBN 978-1-7371679-8-3

Printed in the United States of America

10 9 8 7 6 5 4 3 2 1

www.rationalreasoning.net

To: The College Algebra Student

Welcome!

You are about to begin a new mathematical journey that we hope will lead to your choosing to continue studying mathematics. Even if you don't currently view yourself as a math person, it is very likely that these materials and this course will change your perspective. The materials in this workbook are designed with student learning and success in mind and are based on decades of research on student learning. In addition to becoming more confident in your mathematical abilities, the reasoning patterns, problem solving abilities and content knowledge you acquire will make more advanced courses in mathematics, the sciences, engineering, nursing, and business more accessible. The worksheets and homework will help you see a purpose for learning and understanding the ideas of algebra, while also helping you acquire critical knowledge and ways of thinking that you will need for learning mathematics in the future. To assure your success, we urge you to advantage of the many resources we have provided to support your learning. We also ask that you make a strong effort to make sense of the questions and ideas that you encounter. This will assure that your mathematical journey through this course is rewarding and transformational.

Wishing you much success!

Dr. Marilyn P. Carlson

Table of Contents

This investigation contains review and practice with important skills and procedures you may need in this module and future modules. Your instructor may assign this investigation as an introduction to the module or may ask you to complete select exercises "just in time" to help you when needed. Alternatively, you can complete these exercises on your own to help review important skills.

Order of Operations
Use this section prior to the module or with/after Investigation 1.

The order of operations is an agreed-upon convention that ensures expressions are evaluated the same way by everyone. The commonly accepted order in the U.S. is as follows.

    i. *Simplify any expressions contained within parentheses or brackets. If there are nested sets of parentheses, then work from the inside to the outside.*

    ii. *Evaluate any parts of the expression containing exponents.*

    iii. *Negate any terms with a negation symbol in front of them.*

    iv. *Perform all multiplication and division, working from left to right.*

    v. *Perform all addition and subtraction, working from left to right.*

In Exercises #1-3, use the order of operations to simplify each expression.

1. $3 + 6(4) \div 12 - 1$

2. $4(3 + 2)^2 - 10$

3. $-(16 - 2^3) + 4$

When an expression is written as a fraction, we think of the entire numerator and entire denominator as each having parentheses around them. Thus, we simplify the entire numerator as much as possible and the entire denominator as much as possible before reducing or evaluating the quotient.

In Exercises #4-6, use the order of operations to simplify each expression.

4. $\dfrac{13 - 3^2}{2^3}$

5. $\dfrac{4 - 6 \div 2}{10 + 3 \cdot (-3)}$

6. $\dfrac{-5 + \left(9 - 5(4)\right) + 3}{5(-2) - 1}$

In Exercises #7-9, place parentheses to guide the order of operations so that each expression evaluates to the indicated value. For example, if we want the expression $3 + 6 \cdot 4 - 2$ to evaluate to 18 we need parentheses as follows: $(3 + 6) \cdot (4 - 2)$. Any other way of placing the parentheses, or evaluating it as its written, will produce a different result.

7. $3 + 6 \cdot 4 - 2$ so that its value is 34

8. $5 - 4 \cdot 2^2 + 18$ so that its value is 22

9. $5 - 4 \cdot 2^2 + 18$ so that its value is –41

The Distributive Property
Use this section prior to the module or with/after Investigation 2.

The distributive property is stated as follows. For real numbers a, b, and c,

$a(b + c) = ab + ac$ [can also be written $ab + ac = a(b + c)$]

The distributive property is all about grouping. Suppose six friends each bring two bags of tortilla chips and three bags of pretzels to stock the snack bar at a dorm movie night. How many bags of snacks did they bring?

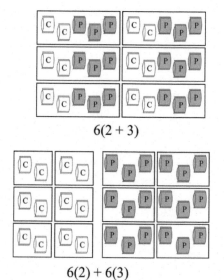

$6(2 + 3)$

$6(2) + 6(3)$

We can think of this in two ways. The first method is to count the number of bags of snacks each person brings [(2 + 3), or 5] and multiply this by the number of people [6]. So the total number of bags of snacks [30] can be represented as shown to the right (top).

The second method is to count the number of bags of tortilla chips [6(2), or 12] and the number of bags of pretzels [6(3), or 18] and add them together [12 + 18]. So the total number of bags of snacks [30] can be represented as shown to the right (bottom).

Therefore, we can understand why the expressions 6(2 + 3) and 6(2) + 6(3) are equivalent (beyond the fact that they evaluate to the same number).

In Exercises #10-12, use the distributive property to rewrite the expression in expanded form. For example, we can rewrite $3(6 + x)$ as $3(6) + 3x$ or $18 + 3x$.

10. $2(5 + r)$

11. $7(3x + 2)$

12. $x(7 + 4y)$

When subtraction is involved, you might choose to change the expression to involve addition. This may help you avoid sign errors (including forgetting about the negative). For example, you can rewrite $2(x - 5)$ as $2(x + (-5))$. Then applying the distributive property yields $2x + 2(-5)$, or $2x + (-10)$, or $2x - 10$.

In Exercises #13-15, use the distributive property to rewrite the expression in expanded form.

13. $6(x - 4)$

14. $4(3 - 2y)$

15. $-x(6y - 5)$

Note that the distributive property works "both directions" (we usually call one direction *distribution* and the other direction *factoring* – but they are both applications of this equality). So just as we can rewrite $3(x + 5)$ as $3(x) + 3(5)$ or $3x + 15$, we can rewrite $3x + 15$ as $3(x + 5)$.

In Exercises #16-21, use the distributive property to rewrite the expression in factored form. For example, we can rewrite $3x + 15$ as $3(x + 5)$.

16. $4x + 36$

17. $2x + 14$

18. $5x - 20$

19. $3x - 27$

20. $-2x + 30$

21. $-6x - 24$

Evaluating Expressions
Use this section prior to the module or with/after Investigation 2.

In Exercises #22-24, evaluate each expression given that $a = 2$, $b = -3$, $c = 0$, and $d = -5$.

22. $b(2a - 3d)$

23. $d^2 - 25 + c$

24. $6 - \dfrac{b}{ad}$

Formulas and Equations

Use this section prior to the module or with/after Investigation 3.

Formulas show how the values of two (or more quantities) are related as they change together. ***Evaluating a formula*** involves substituting value(s) into the expression on one side of the formula (the side that contains operations and perhaps multiple terms) to determine the value of one of the quantities.

In Exercises #25-27, evaluate each formula for the given input.

25. $y = (x+5)^2 - 6$ for $x = -1$ 26. $c = \frac{1}{3}(d-4)+7$ for $d = 12$ 27. $y = \dfrac{1}{x} + x^2$ for $x = 10$

Solving an equation involves fixing the values of both sides of a formula that represents the relationship between two quantities' values as they change together and then determining the value(s) of the remaining variable(s) that produce that value. For example, in the formula $y = 2x + 3$, both y and $2x + 3$ are variable expressions that can take on many different values. To solve the equation for x when $y = 15$ means that we fix the value of y and $2x + 3$ so that they both must be 15, then determine the value(s) of x that make this true, or the value(s) of x such that $2x + 3 = 15$.

In Exercises #28-29, use the given information to write an equation and then solve it.

28. solve $4x + 15 = y$ for x if $y = 38$ 29. solve $z = \dfrac{x+3}{x}$ for x if $z = 7$

Writing a formula involves representing a relationship between two (or more) quantities that change together using mathematical expressions connected by an equal sign.

30. We define the following variables to represent dimensions of a rectangle that can change.

   l = length of the side of the rectangle (measured in inches)
   w = width of the side of the rectangle (measured in inches)
   p = perimeter of the rectangle (measured in inches)
   A = area of the rectangle (measured in square inches)

   a. Write the formula that represents a rectangle's perimeter given its width and length.

   b. Determine a rectangle's width (in inches), w, when $p = 21$ and $l = 7$. Show your work.

This lesson introduces the concepts of quantity and variable, two ideas that are key to defining formulas and graphs that represent how two quantities' values are related and change together (e.g., the volume of water in a fish tank and the depth of the water in the tank).

The first step in constructing formulas and graphs in applied problems is to identify words that describe a quantity—an attribute of some situation or object that can be measured or assigned a value. Next, representing these quantities in a drawing can help you visualize how the quantities are related. Throughout this course you will be asked to write formulas to define one quantity in a problem context "in terms of" another quantity. The process of constructing this formula always begins with you developing a personal image of how the quantities described in the problem statement are related and how their values vary together. Learning to conceptualize and relate quantities' values in a problem will unlock your potential to construct formulas and graphs that are meaningful to you. Consistent use of this approach will give you confidence in confronting word problems and lead to you becoming a powerful mathematical thinker!

---

*1. Arizona State University graduate Desiree Linden won the Boston Marathon in 2018. She began the 26.2-mile race with a slow pace due to wind and rain. After completing the first half of the race, she increased her pace, completing the marathon in a time of 2:39:54 (2 hrs., 39 min., and 54 sec.).

    a. Identify at least two distances in this situation that are constant (that is, their values do not vary). *Be sure to specify the distance, its value, and the unit of measure.*

    b. Identify at least two distances in this situation that vary in value. *For each distance, be sure to specify the unit of measure and where the measurement begins.*

    c. When Desiree has run 6 miles from the starting line of the 26.2-mile race, how far is she from the finish line? Indicate Desiree's approximate position on this illustration.

    d. When Desiree has run 22 miles from the starting line of the 26.2-mile race, how far is she from the finish line? Indicate Desiree's approximate position on this illustration.

    e. As Desiree's distance from the starting line increases, how does her distance from the finish line change?

*2. a. Represent each of the following on the given illustration.
   i) The total race length using a solid line segment with bars at each end.
   ii) Desiree's distance from the starting line using a dashed line segment with a dot on one end, to indicate where the measurement begins (at the starting line) and an arrow on the right end to represent the direction of the measurement.
   iii) Desiree's distance from the finish line, using a dashed line segment with a dot on one end to indicate where the measurement begins (at the finish line) and an arrow on the left end to represent the direction of the measurement.

   b. As Desiree runs the race, how do each of the following vary?
   i) Desiree's distance from the finish line.   ii) Desiree's distance from the starting line.

   iii) The lengths of the line segments you drew in part (a).

   c. In mathematics it is useful to use a letter (referred to as a **variable**) to represent the value of a quantity when that value varies. If we let the variable x represent Desiree's distance from the starting line (in miles), what does the value of the expression $26.2 - x$ represent?

   d. Label each line segment you drew in part (a) with one of the following expressions that represents its value: 26.2, x, or $26.2 - x$.

   e. What values can x assume (take on) in this situation? Give three examples and then describe all possible values.

   f. T or F: The value of $26.2 - x$ varies as Desiree runs the race.

   g. T or F: The expression $26.2 - x$ represents Desiree's distance from the finish line **in terms of** her distance from the starting line, x. Select your answer and then explain what the phrase "in terms of" means in this situation.

h. As the value of x, Desiree's distance from the starting line (in miles), increases from 0 to 26.2, her distance from the finish line <increases or decreases> from _____ to _____. (*Be sure to include the units in your answer.*)

Quantities and Variables

Quantities are the attributes of an object or situation that you can imagine measuring. It is important to be descriptive when identifying a quantity so that the quantities you identify are clearly distinguished (e.g., saying "Desiree's distance from the finish line" instead of "Desiree's distance" or just "distance").

When assigning a *variable* to represent the value of a quantity (when that value varies), we must specify even more details including:
  i. what quantity we are measuring,
  ii. where the measurement begins,
  iii. the direction of the measurement (e.g., above the ground, east of the stop sign), and
  iv. the measurement unit (e.g., inches, seconds).

Using Exercise #1, we might define a variable as follows:

Let x represent Desiree's distance (in miles) from the starting line.

We then think of x as varying in value. As Desiree runs the race her distance (in miles) from the starting line varies.

When a quantity's value can vary in a situation, we say that it is *a quantity that varies in value*. Example: An airplane's altitude (in feet) above 5000 feet of elevation. In this situation, the airplane's *altitude* is a measurable attribute of the airplane. The airplane's altitude is measured from 5000 feet above sea level. The unit of measurement is feet.

When a quantity's value does not vary in a situation, we say that it is *a quantity with a fixed value*. Example: The Washington Monument's height (in feet). In this situation, the Washington Monument's *height* is a measurable attribute of the Washington Monument. The height, measured from the monument's base, does not change. The unit of measurement is feet.

---

3. A tortoise and hare are competing in a race around a 400-meter track. The arrogant hare decides to let the tortoise have a 150-meter head start. When the starting gun is fired, the hare begins running at a constant speed of 6.5 meters per second and the tortoise begins crawling at a constant speed of 2.5 meters per second.

   a. Underline each phrase in the above text that provides information about a quantity.

   b. List each quantity that varies in value in the tortoise and hare context. *Be sure to say where the quantity is being measured from, the direction of measurement, and the unit of measurement.*

c. Make **two** drawings of the situation representing two different moments in time during the race. On each of your drawings, do the following.
    *   Use a solid line segment with bars on the end to represent distances with a fixed value.
    *   Use a dashed line segment with a dot on one end and an arrow on the other end to represent the starting point and direction of the measurement of distances with a varying value.

d. Define a variable to represent the time elapsed since the starting gun was fired. Be specific in your description, making sure you address criteria (i) – (iv) in the definition box for defining a variable.

e. Use the variable you defined in part (d) to write an expression that represents the hare's distance from the starting line (in meters).

f. Use the variable you defined in part (d) to write an expression that represents the tortoise's distance from the starting line (in meters).

g. Use the variable you defined in part (d) to write an expression that represents how many meters the tortoise is ahead of the hare.

h. Who won the 400-meter race? Justify your answer.

---

*4. A ball is thrown upward by a person standing on the ground. If we want to represent *the height of the ball above the top of a 15-foot tall building,* we can assume that the building's height is being measured from the ground and that the ball can be both above and below the top of the 15-foot tall building.

Here is a number line with 0 representing the top of the 15-foot tall building. We labeled the number line so that numbers to the right of 0 represent the ball's height **above** the top of the building (number of feet).

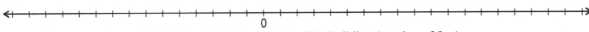

0

the ball's height above the top of the building (number of feet)

a. What do the negative values indicate in this situation?

b. Define a variable d to represent the ball's height above the top of the building. (*Recall the four criteria for defining a variable.*)

c. What does a value of $d = -5$ convey about the height of the ball relative to the top of the building?

d. Illustrate on the given number line the ball's height above the 15-foot building increasing from $d = -6$ feet to $d = 8$ feet. How many feet did the ball travel, and in what direction was the ball traveling?

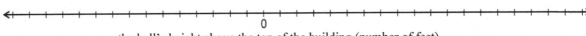

0

the ball's height above the top of the building (number of feet)

*5. a. With a partner or as a class, discuss the usefulness of introducing a variable (symbol) to represent the measurement of a quantity when its value can vary. Make notes about important conclusions you discuss.

---

b. When defining variables, it is important to be specific in describing the quantity with a value represented by the designated symbol.

Is the statement "d = Erin's distance (in miles)" an adequate variable definition? If not, what is missing and how might you modify this statement to improve it? (*You can make assumptions about the situation as needed.*)

c. When defining a variable, why is it important to describe where the quantity is being measured from, the direction of the measurement, <u>and</u> the measurement unit?

6. The city's water tank holds 20,500 gallons of water. After the tank is full, water begins draining from the tank.
   a. Define a variable d to represent the volume of water that has drained from the tank.

   b. Use the variable d and the fixed value of 20,500 gallons the tank can hold to write an expression that represents the number of gallons of water remaining in the tank <u>in terms of d</u>.

   c. What values can d assume (take on) in this situation? Give three examples and then describe all possible values.

   d. As d increases from 0 to 20,500, how does the number of gallons of water remaining in the tank change?

   e. The tank stops draining when there are 40 gallons left in the tank. A hose is turned on to refill the tank. If x represents the amount of water *added* to the tank since the hose was turned on, write an expression to represent the amount of water in the tank <u>in terms of x</u>.

In this investigation, we continue to explore how to use expressions, formulas, and graphs to represent quantities' values and the relationship between two quantities' values as they change together.

*1. At the beginning of the year 1900, the U.S. population was about 76,300,000 and Canada's population was about 5,500,000. Let u represent the U.S. population some number of years since the beginning of 1900, and let c represent the Canadian population the same number of years since the beginning of 1900.

   Based on this context, consider each of the following expressions. What quantity's value is represented in each case?

   a. $u + c$

   b. $u - c$

   c. $\dfrac{u}{c}$

   d. $c - 5{,}500{,}000$

"in terms of"

When we say that one quantity is expressed **"in terms of"** a second quantity, it means we've written an expression to represent values of the first quantity based on performing operations involving values of the second quantity. (*We first encountered this phrase in Investigation 1.*)

For example, let d represent the distance Desiree Linden has run since starting the Boston Marathon (in miles). The Boston Marathon is a 26.2-mile race.

total length of the Boston Marathon (miles)	distance Desiree has run since starting the marathon (miles)
26.2	$- \quad d$

Desiree's distance from the finish line (miles)

Then "$26.2 - d$" represents Desiree's distance from the finish line (in miles) **in terms of** the distance she has run since the race started. This is because her distance from the finish line is represented with an expression involving the value of the distance she has run since the race started, d.

Note that the value of $26.2 - d$ varies with the value of x.

---

*2. Given the following variable definitions, complete the tasks in this exercise.
   x = the length of the side of a square (in inches)
   p = the square's perimeter length (in inches)
   A = the square's area (in square inches)

   a. Write an expression that represents the square's perimeter length in terms of its side length.

   b. What is the square's perimeter length when its sides are 5.9 inches long?

---

c. Write an expression that represents the square's side length in terms of its perimeter length.

d. Write an expression that represents the square's side length in terms of its area.

e. Write an expression that represents the square's area in terms of its perimeter length.

f. What is the square's area if its perimeter is 42.8 inches long?

g. Write an expression to represent a rectangle's width in terms of its perimeter length and the rectangle's length. Define variables and make a drawing to help you visualize the quantities in the problem context and how they are related. Assume lengths will be measured in inches.

Representing Relationships on a Coordinate Plane

A ***coordinate plane*** consists of a vertical axis and a horizontal axis that intersect at a point. The intersection point is called the ***origin*** and is labeled (0, 0).

When we express the value of Quantity B *in terms of* the value of Quantity A using a graph, we will use the following mathematical conventions.

- We represent values of Quantity B on the vertical axis and values of Quantity A on the horizontal axis.
- We describe points on the plane in the form (a, b) where a represents the value of Quantity A and b represents the value of Quantity B.
- We call the value of Quantity A the *independent value* and the value of Quantity B the *dependent value*.

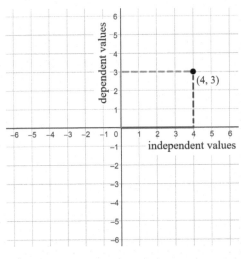

3. a. T or F: The vertical axis represents values of the independent variable.

   b. T or F: The point (4, 3) on a graph that relates the value of Quantity B in terms of the value of Quantity A conveys that, when the value of Quantity A is 4, the value of Quantity B is 3.

*4. Hector and Brandon are 45 feet apart and start walking toward one another at the same time. Hector is walking twice as fast as Brandon, so when Brandon travels some number of feet, Hector travels two times as far. The two men stop when they reach each other.

   a. Write a formula that expresses the distance between Hector and Brandon ***in terms of*** the number of feet Brandon has walked. Define any variables you use.

---

b. Label the axes on the given coordinate plane based on the conventions described in the definition box.

c. Plot the points (0, 45), (10, 15), and (15, 0) on the coordinate plane. What do each of these points represent in this context?

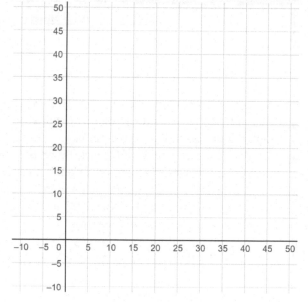

d. Draw a line through the points you plotted. What do the points that make up this line represent?

e. On the coordinate plane, represent the distance Brandon has walked increasing from 0 to 5 feet.

f. As the distance Brandon has walked increases from 0 to 5 feet, how does the distance between Brandon and Hector change? Answer this question, and then represent this change on the graph.

Changes in Quantities

Once we have defined a variable (such as x) to represent the value of a varying quantity, we can write Δx (read as "change in x" or "delta x") to represent *changes* in that quantity's value.

The change in a quantity's value is a new quantity itself. If the value of x changes from x_1 to x_2, then the amount of change can be represented as $x_2 - x_1$.

For example, if x changes from $x = 6$ to $x = 19$, then $\Delta x = 19 - 6$, or $\Delta x = 13$, meaning that

---

5. The temperature in Phoenix, T, in degrees Fahrenheit varies throughout the day. On May 1st the temperature was 65 degrees Fahrenheit at 6 am and 90 degrees Fahrenheit at 11 am.

   a. What was the change in temperature, ΔT, over the period from 6 am to 11 am?

   $25°F$

---

b. Represent the change in temperature, ΔT, determined in part (a) on the number line.

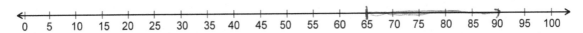

c. Represent a change of temperature of 5 degrees at some point during May 1st.

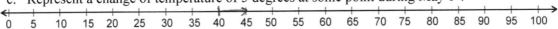

d. How many different ways could you represent the answer to part (c)?

e. Represent a change of –25 degrees Fahrenheit starting from 80 degrees Fahrenheit.

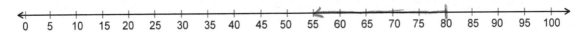

*6. The given graph represents the depth of water in a reservoir in feet in terms of the number of months since January 1, 1990.

a. Define variables to represent the values of the quantities in this situation that can vary.

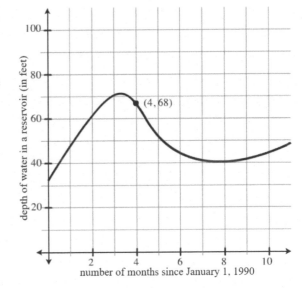

b. Interpret the meaning of the point (4, 68).

c. Label the point (6, 20) on the plane. What does this point represent?

d. Use the graph to estimate the number(s) of months since January 1, 1990 when the depth of water in the reservoir was 45 feet.

e. The number of months since January 1, 1990 increases from 1 to 3 months.
   i. What is the *change* in the number of months since January 1, 1990 over this interval? Use Δ notation to represent your answer.

   ii. Represent the change in part (i) on the graph.

   iii. What is the change in the depth of the water as the number of months since January 1, 1990 varies from 1 to 3 months? Use Δ notation to represent your answer.

   iv. Represent the change in part (iii) on the graph.

f. How does the depth of the water in the reservoir vary as the number of months since January 1, 1990 increases from 4 to 7 months? Illustrate this variation in the depth of the water in the reservoir on the graph.

You have been studying linear relationships and the idea of constant rate of change since your first course in algebra. You learned that any formula that can be written in the form $y = mx + b$ describes a linear relationship between x and y in which x and y vary together at the constant rate of change, m, and that $(0, b)$ represents the *y-intercept or vertical intercept* (that is, $y = b$ when $x = 0$).

In this lesson, we will explore in more depth what it means for two quantities' values to change together at a constant rate of change. **When saying that x and y are changing together at a constant rate of change, m, what restriction does this place on how x and y are changing together?**

*1. You are riding a stationary bike. After your 2-minute warm up, you begin pedaling at a constant rate of 4 minutes per mile for the next 5 miles.

   a. What quantities have values that are changing together in this context?

   b. Discuss with your classmates what it means to be traveling at the constant rate of 4 minutes per mile. (*Say more than, "You travel for 4 minutes every time you pedal a distance of 1 mile," or that "your speed doesn't change." Try to think about a description that will apply no matter what interval of distance we look at.*) Make note of any important conclusions.

   c. When riding at a constant rate of 4 minutes/mile:
     i. How many miles would you pedal in 1 minute? Demonstrate your thinking or discuss your reasoning with a partner.

     ii. How many miles would you pedal in 6 minutes? Demonstrate your thinking or discuss your reasoning with a partner.

     iii. How many miles would you pedal in 9 minutes? Demonstrate your thinking or discuss your reasoning with a partner.

     iv. How many miles would you pedal in ½ minute? Demonstrate your thinking or discuss your reasoning with a partner.

d. Define a variable t to represent your total elapsed time while riding and a variable d to represent your distance pedaled *since you started pedaling at a constant rate*.

e. What do values of Δt and Δd represent in this situation?

f. What generalizations can you make about how Δt and Δd are related when you are riding at a constant rate of 4 minutes per mile?

*2. *Continue with the same context from Exercise #1*: You are riding a stationary bike. After your 2-minute warm up, you begin pedaling at a constant rate of 4 minutes per mile for the next 5 miles.

a. When pedaling at a constant rate of 4 minutes per mile:

i. How many minutes will it take you to pedal $\frac{1}{2}$ mile? Demonstrate your thinking or discuss your reasoning with a partner.

> 2   because $\frac{1}{2} \cdot 4 = 2$

ii. How many minutes will it take you to pedal $\frac{1}{10}$ mile? Demonstrate your thinking or discuss your reasoning with a partner.

> 0.4 minutes   24 seconds
> $4 \cdot 0.1 = 0.4$
> $0.4 \cdot 60 = 24$
>
> $4 \cdot 60 = 240$
> $240 \cdot 0.1 = 24$
> 24 seconds

iii. How many minutes will it take you to pedal 3.2 miles? Demonstrate your thinking or discuss your reasoning with a partner.

> $4 \cdot 3.2 = 12.8$
> 12.8 minutes OR 12 minutes and 48 seconds

d = distance pedaled after warm up (miles)
t = time since start of workout (min)

b. We will create a graph to represent the total amount of <u>time spent riding in terms of the distance</u> pedaled since beginning to pedal at a constant rate. Label the axes accordingly (including units), then complete each part.

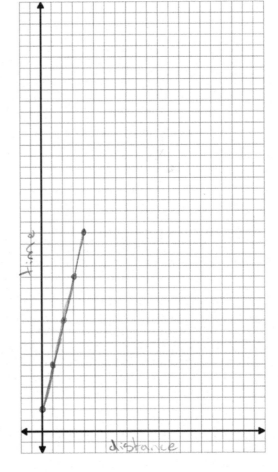

i. Using the coordinate plane, plot the point $(0, 2)$ on the axes and describe what this point represents in the context of this situation.

No distance

2 minutes

ii. Use your work in Exercise #1 to plot 3 more points on your graph.

iii. Draw a line through the points. What do the points that make up this line represent?

iv. Illustrate on your graph an increase in d from 1 to 4 miles. Determine the corresponding value of Δt and illustrate this change on your graph.

v. Complete this sentence: When pedaling at a constant rate of 4 minutes/mile, any change of 1.5 miles will result in a change of _____6_____ minutes.

c. Write a formula that represents the change in the total number of minutes riding, Δt, in terms of the change in the number of miles pedaled (while pedaling at a constant rate), Δd. (*Hint: A change in your riding time is always how many times as large as a change in your distance pedaled?*)

d. Write a formula that represents the total number of minutes you have been riding, t, in terms of the number of miles, d, since you started pedaling at a constant rate of 4 minutes/per mile. Which variable is the independent variable in this formula?

3. The owner of a swimming pool wants to know the rate at which the depth of water in his pool is changing. When he begins filling the pool at 12:00 p.m., there is already some water in the pool. At 2:00 p.m., the depth of the water is 21 inches, and when it is 5:00 p.m., the depth of the water is 32 inches. *See the given table.*

hours since starting to fill the pool	depth of water in the pool (inches)
0	
2	21
3	
5	32
8	
13	

a. Assume that the depth of the water in the pool changes at a constant rate of change with respect to time elapsed. What does this mean? Provide as much detail as possible.

b. What is the change in the time elapsed from 2 p.m. to 5 p.m.? What is the change in the depth of water in the pool over that time period?

c. Assuming a constant rate of change, how fast is the depth of the water in the pool changing (what is the rate of change of the depth of the water in the pool (in inches) with respect to the number of hours since the pool began filling)?

d. Assuming the constant rate of change you found in part (c) holds, complete the given table of values. Discuss (with a partner or as a class) the reasoning you used to complete the table and note any important observations.

e. Make two drawings of the situation at different moments in time to help you visualize how the depth of the water in the pool changes with the time (in hours) since the pool started filling. Be sure to indicate how much water was in the pool when it began filling.

f. Create a formula to represent the depth of the water in the pool in terms of the number of hours since the pool started filling. Define any variables you use.

g. Construct a graph that represents the depth of the water in the pool (in inches) in terms of the amount of time (in hours) since the pool began filling.

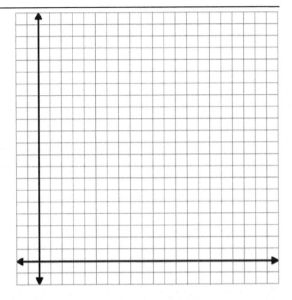

Constant Rate of Change

If x and y are the values of two quantities related by a constant rate of change, then over any interval from x_1 to x_2, where y changes from y_1 to y_2, the following will be true.

- $\Delta x = x_2 - x_1$ represents the change in x's value.

- $\Delta y = y_2 - y_1$ represents the change in y's value.

- The ratio $\dfrac{\Delta y}{\Delta x} = \dfrac{y_2 - y_1}{x_2 - x_1}$ will always have the same value. We use m to represent this value, and we call it the ***constant rate of change of y with respect to x.***

---

*4. Porter received a remote-controlled car as a gift for his birthday. He set the car against a wall and then started it moving straight away from the wall. Porter's car began traveling at top speed 6 meters later and continued at this speed for 15 more seconds. When traveling at top speed, his car traveled 7 meters every 2 seconds.

a. What was the top speed of Porter's car?

b. One of your classmates (Keeley) noted that, "While moving at its top speed, the car travels 7 meters every 2 seconds. We can double each of these values to say that it will travel 14 meters every 4 seconds." Another classmate (Garrett) said, "While moving at its top speed, the car's change in distance (in meters) is always 3.5 times as large as the change in the time elapsed (in seconds). So, in 4 seconds, the car will travel 3.5(4) = 14 meters."

With a partner or as a class, discuss the connections between the reasoning that each student used and why both are valid. Make notes of any observations you make.

c. Construct a formula to represent the distance (in meters) from the wall to Porter's car, d, in terms of the number of seconds *since Porter's car reached top speed, t.*

Constant Rate of Change Revisited

Since $\dfrac{\Delta y}{\Delta x} = m$ represents a constant value when y changes at a constant rate with respect to x, we can rewrite the statement to show that $\Delta y = m \cdot \Delta x$ must also be true. The statement $\Delta y = m \cdot \Delta x$ says that "the change in y is always m times as large as the change in x."

Different contexts might make it easier to reason about the relationship in the form $\dfrac{\Delta y}{\Delta x} = m$ or the form $\Delta y = m \cdot \Delta x$.

---

*5. a. Given that x and y are changing together at a constant rate of change so that $\Delta y = -2 \cdot \Delta x$ and when $x = 1$, $y = 7$, construct the graph of y in terms of x.

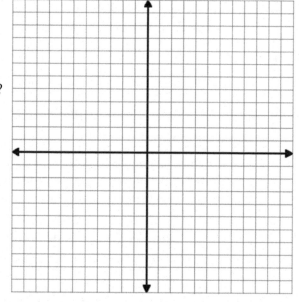

b. As x increases from 1 to 5.5, how does y change? Illustrate these changes on your graph.

c. Does x take on all values between 1 and 5.5? If so, how is this illustrated on your graph?

6. The given graph represents a linear relationship between y, Bob's distance (in miles) to the finish line of a race, and x, the amount of time (in minutes) since Bob passed the final water station.

a. What does the point (–22, 8.2) represent in this situation?

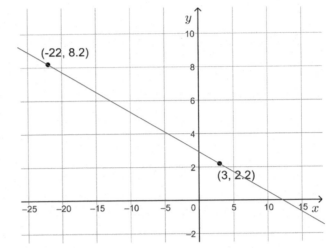

b. What does the point (3, 2.2) represent in this situation?

c. Determine the value of m, the constant rate of change of y with respect to x, in this linear relationship and say what this value represents.

d. Write the formula representing y in terms of x.

*7. Given that the *x*- and *y*-axes have the same scale, answer the following questions for both graphs.

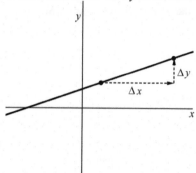

Graph of linear function A

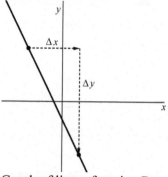

Graph of linear function B

a. What is the approximate constant rate of change of *y* with respect to *x* for each linear function?

Linear function A: Linear function B:

b. Write a formula that expresses Δy in terms of Δx.

Linear function A: Linear function B:

c. Write a formula that expresses *y* in terms of *x* given the *y*-intercept of the graph of function A is (0, 3) and the *y*-intercept of the graph of function B is (0, –3).

Formula for linear function A: Formula for linear function B:

This investigation examines the special case of constant rate of change situations where the following two facts are true.

- There exists a constant rate of change (changes in the two quantities are proportional).
- The quantities *themselves* are proportional.

Mixing two or more substances together in specific amounts is important in many situations. For example, doctors and nurses might dilute liquid medications, gardeners might buy concentrated weed killer that needs to be mixed with water in specific ratios to pour into a sprayer, and bakers need specific amounts of flour, water, sugar, and salt in their recipes based on the batch size. In each of these cases, *proportional reasoning* is key to obtaining correct mixtures.

*1. A nurse has liquid medicine that must be diluted precisely before giving it to a child. The nurse mixes 2 ounces of concentrated medicine with 3 ounces of water based on the manufacturer's instructions.

   a. The circle marked on the cylinder represents the height to which the 5-ounce mixture rises on the cylinder. Shade and label the drawing to indicate the part of the 5-ounce mixture that is the concentrated medicine.

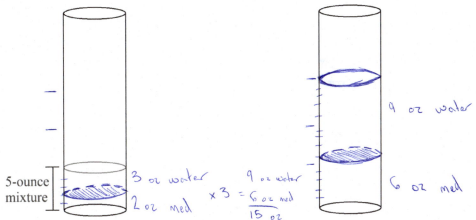

   b. What if the nurse needs to make a 15-ounce mixture? On the empty cylinder, mark the height of the 15-ounce mixture (relative to the height of the 5-ounce mixture). Shade and label the cylinder to indicate the part of the 15-ounce mixture that is concentrated medicine.

   c. For every 6 ounces of concentrated medicine, the mixture contains _____ounces of water. This is because 6 ounces of medicine is 3 times as much as 2 ounces of medicine, so the amount of water will need to be _____times as much as_____ ounces of water.

   d. T or F: The amount of water needed for 3 ounces of medicine is 1.5 times as large as 2 ounces of medicine, so the amount of water needed (according to the directions) is 1.5 times as large as 3 ounces of water (or 4.5 ounces of water).

   e. After mixing the original 2 ounces of medicine and 3 ounces of water, the nurse is directed to give the patient 8 ounces of the mixture instead. You see the nurse adding 3 more ounces to the 5-ounce mixture by (i) adding 1.5 ounces of concentrated medicine and (ii) adding 1.5 ounces of water. Is the patient's dosage:

        (I) accurate   (II) too strong   (III) too weak

   Explain your reasoning.

*2. *Continue with the context from Exercise #1:* A nurse has liquid medicine that must be diluted precisely before giving it to a child. The nurse mixes 2 ounces of concentrated medicine with 3 ounces of water based on the manufacturer's instructions.

a. Complete the table of values for this situation.

volume of concentrated medicine needed (ounces)	volume of water needed (ounces)
1	1.5
2	3
5	7.5
12	18
17	~~22.5~~ 25.5

b. Write instructions for the nurse so they know how much water to add when they know the amount of concentrated medicine they need to use.

c. Write a formula for determining the volume of water to add to *x* ounces of concentrated medicine. (Define your variables before writing your formula.)

$x = oz$ of med
$w = oz$ of water

$w = 1.5x$

d. Create a graph that represents the volume of water needed in terms of the volume of concentrated medicine needed for a proper dilution.

e. Explain what points on your graph represent in this context.

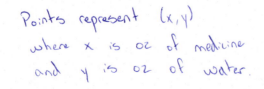

Points represent (x,y) where x is oz of medicine and y is oz of water.

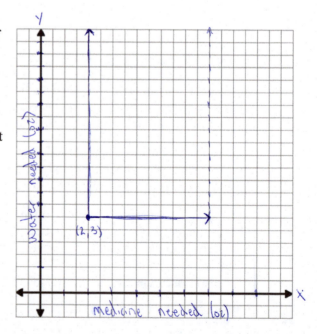

f. Pick a point on the coordinate plane that is ***not*** on the curve/line you drew. Justify why this point is not included on the curve/line.

g. Illustrate the following on the graph you drew.
   i. The point that represents 2 ounces of concentrated medicine and 3 ounces of water.

   ii. An arrow that represents an increase of 5 ounces of concentrated medicine (from 2 ounces) and an arrow that represents the corresponding change in the volume of water. How much did the volume of water change?

   iii. Will a change of 5 ounces of concentrated medicine (from any given volume) always require the same change in the volume of water to maintain the proper dilution? Why or why not? How can you see this on the graph?

   Yes, because otherwise the dilution would be changing.
   From any other value on the graph, a 5 oz change in medicine results in a 1.5 change in water.

h. Fill in the blank: For any volume of concentrated medicine in the mixture, the volume of water in the mixture is always __1.5__ times as large.

   $w = 1.5x$

i. Fill in the blank: For any *change* in the volume of concentrated medicine in the mixture, the *change* in the volume of water in the mixture is always __1.5__ times as large. How can we see this on the graph?

   For every valid point, y is 1.5 times the value of x.
   $\Delta w = 1.5 \Delta x$

Constant rate of change can be ~~x~~ proportional?

Proportional Relationships

In a ***proportional relationship***, two quantities' values are changing so that they remain the same relative size.

One feature of proportional relationships is that the two related quantities must scale together to maintain the proportional relationship. For example, if you multiply the size of one quantity by some number, you must multiply the size of the other quantity by the same number to maintain the relationship.

Continuing with the baking example, if the recipe calls for 5 cups of flour and 3 cups of sugar, and you quadruple the number of cups of flour (to 20), you must also quadruple the number of cups of sugar (to 12). When you do, the number of cups of flour stays 5/3 times as large as the number of cups of sugar.

---

*3. Here are two cylinders: one wide and one narrow. Both cylinders have equally spaced marks for measurement. Water is poured into the wide cylinder up to the 2nd mark. This same volume of water rises to the 7th mark when poured into the narrow cylinder.

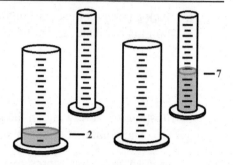

a. Both cylinders are emptied, and water is poured into the narrow cylinder up to the 13th mark. How high would this water rise if it were poured into the empty wide cylinder?

b. Imagine pouring water into the narrow cylinder, and let x represent the height (mark-number) the water reaches in the narrow cylinder as you pour. Let y represent the height (mark-number) the same volume of water would reach if poured into the wide cylinder.

   i. What relationship is conveyed by the statement $\frac{y}{x} = \frac{2}{7}$ in this context? (*Hint: What quantity's value is represented on each side of the equation?*)

   ii. What relationship is conveyed by the statement $y = \frac{2}{7}x$ in this context? (*Hint: What quantity's value is represented on each side of the equation?*)

   iii. What relationship is conveyed by the statement $\Delta y = \frac{2}{7} \cdot \Delta x$ in this context? (*Hint: What quantity's value is represented on each side of the equation?*)

c. When $x = 10$, what is the value of y? What does this value tell us?

d. Compare and contrast what the three statements, $\frac{y}{x} = \frac{2}{7}$, $y = \frac{2}{7}x$, and $\Delta y = \frac{2}{7} \cdot \Delta x$ convey about how the water heights and changes in water heights for the two cylinders are constrained and related.

4. A 19-inch candle is lit. The candle burns away at a constant rate of 2.5 inches per hour.
   a. Draw two "snapshots" of the candle as it is burning. Represent each quantity with a fixed measurement using a solid line segment and each quantity that varies in value using a dashed line segment and arrow.

   b. Define an ***expression*** to represent the candle's remaining length (in inches) in terms of the number of hours since the candle started burning, t. Use this expression to label any line segments representing this length in your diagrams in part (a).

   c. Define a ***formula*** to represent the remaining candle length (in inches), r, in terms of the number of hours since the candle started burning, t.

   d. Use a formula to represent the ***change*** in the remaining candle length (in inches), Δr, in terms of the change in the number of hours the candle has been burning, Δt.

   e. Which quantities in this situation are proportional? Explain.

   f. T or F: Since the change in the remaining candle length, Δr, is proportional to the change in the number of hours the candle has been burning, Δt, this means that the remaining candle length (in inches), r, is proportional to the number of hours the candle has been burning since it was first lit, t. Justify your answer.

g. How much time will it take for the candle's remaining length to become 0 (that is, after how long will the entire candle have burned away)? Show your work and explain your reasoning using the proportional relationship that constrains how Δr and Δt change together.

h. Discuss with a partner or as a class how the idea of **proportional quantities** is related to and different from the idea of **constant rate of change of one quantity with respect to another**. Make note of any key conclusions.

*5. A 17-inch candle is lit and burns at a constant rate of 1.8 inches per hour. Let t represent the number of hours since the candle started burning. Let r represent the remaining candle length (in inches).

a. What does −1.8 represent in this context? Include units in your answer.

b. What does the expression $1.8t$ represent in this context?

c. What does the expression $17 - 1.8t$ represent in this context?

d. Is Δr proportional to Δt? Justify your answer.

e. Is r proportional to t? Justify your answer.

*1. Laura has just won tickets to a concert that begins in 3 hours. She immediately calls her friend Haley, who lives in a city 282 miles from the location of the concert, and invites her to meet her at the concert.

    a. Discuss and decide what assumptions and/or ideas might be useful to decide if Haley should start driving to the location of the concert?

    b. Determine the *average speed* Haley would need to drive to make it to the concert on time. What does this *average speed* represent in the context of this problem? Is it feasible for Haley to make it to the concert on time?

*2. Isabel and Miles are driving to a wedding 145 miles from their home. They left their house 3 hours before the start of the wedding. Due to icy road conditions, they were only able to drive 90 miles in the first 2 hours since they left their home.

    a. If they maintain the same average speed as they continue driving on the same icy roads, will they make it to the wedding on time? Justify your answer with a mathematical argument.

    b. If the road conditions improve, at what "average speed" will they need to drive to make it to the wedding on time?

    c. What does the average speed in part (b) represent in the context of this problem?

3. A policeman is positioned on the side of a road *without* a radar gun. What information will the officer need to determine if a car is speeding along the road? Explain.

*4. A car is driving away from a crosswalk. The distance, d (in feet), of the car north of the crosswalk t seconds since the car started moving is given by the formula $d = t^2 + 3.5$.

   a. As the number of seconds since the car started moving increases from 1 second to 4 seconds, what is the change in the car's distance north of the crosswalk? (*Use the given formula to help determine this change in distance.*)

$$d = 1^2 + 3.5 \qquad d = 4^2 + 3.5 \qquad \Delta d = 19.5 - 4.5$$
$$= 4.5 \qquad\qquad = 16 + 3.5 \qquad\qquad = 15$$
$$\qquad\qquad\qquad = 19.5$$

   b. Label the axes and construct a graph that represents the distance (in feet) of the car north of the crosswalk, d, in terms of t, the number of seconds since the car started moving. Plot at least 3 points and then sketch the curve.

   c. What do the points that make up the curve represent?

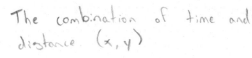
The combination of time and distance. (x, y)

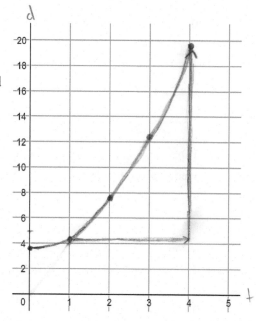

   d. Illustrate each of the following on your graph.

   i. An increase in t from 1 to 4 seconds.

   ii. The corresponding change in the car's distance north of the crosswalk, as the value of t increased from 1 to 4 seconds.

e. True or False: The car travels at a constant speed as the value of *t* increases from 1 to 4 seconds. Discuss and explain. (*Hint: It may help to think about how far the car travels in the 1st second of the interval as compared to how far it travels in the 2nd second of the interval*).

False

It increases exponentially.

f. Since the car is not traveling at a constant speed, it can be challenging to describe the car's speed. It is a common practice to <u>estimate</u> the speed of an object over an interval of time by pretending its speed is constant, and asking, "*At what constant speed would the object have to travel in order to travel the same distance in the same amount of time?*"

For this problem context, answer the following questions to determine the constant speed needed to travel the same distance that the car actually traveled over the time interval *t* = 1 to *t* = 4 seconds.

i. Plot the points (1, 4.5) and (4, 19.5) on the graph you created in part (b) and draw a straight line passing through them.

ii. Determine the constant rate of change of the linear relationship (slope of the line) represented by the line you drew. (*Be sure to include units in your answer.*)

g. True or False: The points making up the line you drew represent the actual distance (in feet) of the car north of the crosswalk for various values of *t*, as *t* increases from 1 to 4 seconds. Explain.

*5. The distance, d (in feet), of a car south of an intersection t seconds after it started moving is given by the formula $d = 2t^2 - 8t + 8$. [*Note: We will use subscripts to refer to different values of d and t.*]

   a. Determine the value of d_1 when $t_1 = 2$, and describe what the values of d_1 and t_1 represent. (*Hint: When 2 seconds have passed since..., the car is...*).

   b. Determine the value of d_2 when $t_2 = 5$, and describe what the values of d_2 and t_2 represent.

   c. Determine the value of $\dfrac{d_2 - d_1}{t_2 - t_1}$ and explain what this value represents in the context of this situation. (*Hint: What constant rate of change is this?*)

   d. Label the axes and construct a graph of $d = 2t^2 - 8t + 8$ for values of $t \geq 0$. (*You may use a graphing calculator or set up a table.*)

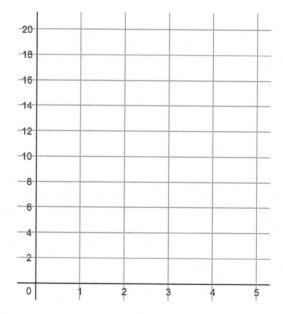

   e. Describe how the car's distance south of the intersection varies over the first 5 seconds since it started moving.

   f. On your graph, illustrate the constant speed another car would need to travel in order to cover the same distance (as the car actually traveled) over the time interval from $t_1 = 2$ to $t_2 = 5$.

g. If the car continued traveling at a constant speed of 6 feet per second on the interval from $t = 5$ to $t = 13$ seconds since the car started moving, how far would the car be south of the stop sign 13 seconds after it started moving?

Average Rate of Change (over some interval)

The *average rate of change* of a function over some interval of the domain is the constant rate of change over that interval, $[x_1, x_2]$, that produces the same net change, $y_2 - y_1$, in the function's output value on that interval.

Graphically, the *average rate of change* of a function on an interval of a function's domain $x_1 \leq x \leq x_2$ is represented as the slope of the line passing through the points (x_1, y_1) and (x_2, y_2). The average rate of change can be calculated using the slope formula $m = \frac{y_2 - y_1}{x_2 - x_1}$.

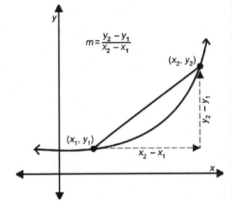

The average rate of change is a common tool for estimating the rate of change of a function over some interval of the function's domain in which the rate of change is not constant.

You have likely observed that determining an average speed of some moving object on a time interval from t_1 to t_2 involves determining the constant speed needed to travel the same distance that the object actually traveled over the specified time interval (from t_1 to t_2).

*6. Given that a car's distance (in feet) east of an intersection t seconds since it started moving is represented by $d = t^2 - 2t$, determine the car's average speed over each interval of elapsed time. ***Be sure to include the units in your answer!***

a. $t_1 = 2$ to $t_2 = 6$ b. $t_2 = 6$ to $t_3 = 10$ c. $t_3 = 10$ to $t_4 = 11$

d. What observations can you make about how the car's speed is changing on the time interval from $t_1 = 2$ to $t_4 = 11$ since the car started moving east of the intersection?

*1. You are considering renting a house located at the intersection of D Street and 23rd Avenue. The closest cell tower for your cell phone provider is at the intersection of B Street and 20th Avenue. Each street is located one mile apart. If you are farther than 3.5 miles away from the cell tower, you will not have good reception.

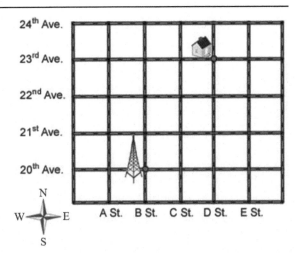

a. What quantity do you need to find to determine if you will have good reception at the house? What known quantities in the context will you use to determine the direct distance from the house and the cell tower? Illustrate these quantities on the given figure.

b. Determine the distance between the cell tower and the house.

c. Let the bottom left corner of the grid be defined as location $(0, 0)$. If another cell tower's position on the grid is at $(6, 2)$ and the house's position is at $(4, 4)$, illustrate what the value of the expression $6 - 4$ represents on the grid.

d. Use the two points to determine the east-west and north-south distances from the second cell tower to the house and then use these values to determine the straight-line distance between this second cell tower and the house.

e. If the point (x, y) represents the position of any house on the grid, explain what $x - 2$ and $y - 1$ represent in this problem context.

f. Use the expressions in part (e) to define an expression that represents the distance of *any house*, located at (x, y), from the cell tower located at (2, 1).

g. A new tower is scheduled to be built at some location on the grid, (h_1, v_1) (its exact location will be determined at a later date). Define a formula to determine the distance of any house, d (in miles), from this third cell tower if the house is located at (x, y).

*2. Consider a cell tower at the location (–2, 4) on the given grid with a coverage radius of 3 miles. *The positive vertical axis points north and the positive horizontal axis points east.*

    Every point 3 miles away from the cell tower will be on the boundary of the cell coverage. This will form a circle around the point (–2, 4) with a radius 3 units long.

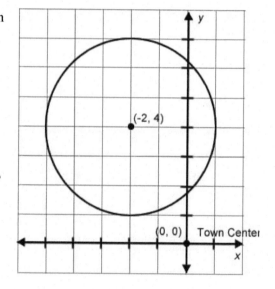

a. A point $(x, 2)$ southeast of the tower is on the service boundary. What must be the value of x?

b. Let (x, y) be any point on the service boundary. Create a formula that represents the distance (in miles) of any point on the service boundary from the cell tower located at (–2, 4).

*1. A carpenter needs to cut a wooden board to a length of 7.6 inches with a **tolerance** of 0.1 inch, meaning the actual length after the cut can be within 0.1 inch of 7.6 inches and still be usable.

   a. Define a variable to represent the length of the board after cutting. *Be sure to include units.*

   b. List some possible values that represent usable board lengths for the variable you defined in part (a), and then represent all possible values of usable board lengths on the given number line.

   c. Write a mathematical statement (inequality) using the variable you defined in part (a) to represent the possible usable board lengths.

   d. We will call the difference between the actual board length and the desired board length the ***error***. Write an expression using the variable you defined in part (a) to represent the possible error for a usable board length.

   e. Write a mathematical statement (inequality) using the variable you defined in part (a) that represents errors within our tolerance of 0.1 inch.

   f. Write a mathematical statement (inequality) using the variable you defined in part (a) that represents actual board lengths that are **<u>not</u>** usable, then represent these values on a number line.

*2. Let x represent the value of a quantity. We want to choose x-values within 4.5 units of 16.

   a. List some values of x that meet this constraint.

   b. Write a mathematical statement (inequality) to describe the possible changes in x away from 16 that meet the given constraint.

c. Write a mathematical statement (inequality) to describe the possible values of x that meet the given constraint and then represent these values on the given number line.

3. Let x represent the value of a quantity. We want changes in the value of x away from 5 that are positive but no more than 3 units.

   a. List some values of x that meet this constraint.

   b. Write a mathematical statement (inequality) to describe the possible changes in x away from 5 that meet the given constraint.

   c. Write a mathematical statement (inequality) to describe the possible values of x that meet the given constraint and then represent these values on the given number line.

4. Let x represent the value of a quantity. We want to choose values of x that are at least 4 units away from $x = -3$.

   a. List some values of x that are at least 4 units away from $x = -3$, then represent all possible values that meet this constraint on the number line.

   b. Write a mathematical statement (inequality) that describes values of x at least 4 units away from $x = -3$. Try to think of at least two different ways to write this.

<u>**Absolute Value**</u>

Given real numbers h and c with $c > 0$ and the real number variable x, the expression $|x - h|$ represents the magnitude of the difference between x and h. When represented on a number line, it is the distance between x and h. Therefore,

- $|x - h| = c$ represents that the difference between x and h is exactly c units, or that x and h are exactly c units apart on a number line.

- $|x - h| < c$ represents that the difference between x and h is less than c units, or that x and h are less than c units apart on the number line.

- $|x - h| > c$ represents that the difference between x and h is more than c units, or that x and h are more than c units apart on the number line.

The solutions to these statements are the values of x that meet the given constraint. For example, $|x - 11| \leq 3$ could be interpreted as saying that the distance from 11 to x is less than or equal to 3 units. What values of x are at most 3 units away from 11? The answer would be all values of x between 8 and 14, which we can represent as $8 \leq x \leq 14$.

5. Use absolute value notation to rewrite your solutions to Exercises #1e, #2b, and #4b. (If your answers are already expressed using absolute value notation you can skip this question.)

*6. Jamie is assembling a model airplane that flies. The instructions say that the best final weight is 28.5 ounces. However, the actual weight can vary from this by up to 0.6 ounces without affecting how well the model flies.
   a. (i) Define a variable to represent the weight of the assembled model airplane, (ii) represent the weights that are "acceptable" on the number line, and then (iii) describe them using an inequality.

   b. Write a mathematical statement (inequality) that represents the acceptable variation in the model airplane's weight…
      i. using absolute value.   ii. without using absolute value.

---

7. a. What do the solutions to the inequality $|x-5| < 2$ represent?

    b. Represent all values of x that make the statement true on the number line and as an inequality.

    c. Rewrite the statement in part (a) without using absolute value.

8. a. Given that we want $(w-1) \geq 5$ or $(w-1) \leq -5$, what do the solution values represent?

    b. Represent all values of w that make the statement true on the number line and using inequalities.

    c. Rewrite the statement in part (a) using absolute value.

*9. Use absolute value notation to represent the following.
  a. x differs from 6 by no more than 7.5 units.

  b. The distance from -2 to r on a number line is more than 1 unit.

*10. For each statement, do the following.
  i.   Describe the meaning of the statement relative to values on a number line.
  ii. Rewrite the statement without using absolute value.
  iii. Determine the values of the variable that make each statement true.
  iv. Represent the solutions on a number line.

  a. $|x-11|<6$ b. $|y+7|\geq1$

  c. $|a-3.3|>0.8$ d. $|p+2|\leq5$

Introducing Rate of Change from a Co-variation of Quantities Perspective

---

I. QUANTITIES AND CO-VARIATION OF QUANTITIES (Text: S1)

In Exercises #1-2, you are given a situation. For each situation, do the following.
- (i) Identify at least one quantity with a fixed value and give a reasonable unit of measure.
- (ii) Identify at least two quantities that vary in value and give a reasonable unit of measure for each. Define variables to represent the values of these quantities.

1. a. A mountain climber hikes with two friends for 5 hours.
    b. A laptop computer charges for 8 hours each night.
    c. A student studies for 6 hours each weekend.
    d. Jessica bikes 30 miles around a 5-mile course.

2. a. A 10-inch candle burns for 2 hours.
    b. A stone layer installs square tiles that have a 9-inch diagonal length on patio floors.
    c. A girl runs around a ¼-mile track.
    d. A scuba diver descends from the surface of the water to a depth of 60 feet.

3. The time allotted for students to complete an exam varies with the number of questions on the exam. For each question there are 5 minutes allotted. Write a formula that represents the number of minutes allotted to take an exam based on the number of questions on the exam. (*Define any variables you use.*)

4. There are twelve times as many football players on a football team as there are coaches. Write a formula that represents the number of coaches on a team based on the number of players on the team. (*Define any variables you use*).

5. Write an equation for each of the given relationships. Then solve the equation to determine the value of the unknown number.
    a. 17.5 is equal to 2 times some number. What is the number?
    b. The sum of 3 times some number and 12 is 42. What is the number?
    c. ¼ of some number is 4.3. What is the number?
    d. Some number is 4 times as large as 9.8. What is the number?
    e. Some number is equal to 1/3 of the sum of 88.2, 93.5, and 64. What is the number?
    f. The change from some number to 12.5 is –5. What is the number?

6. Write an equation for each of the given relationships. Then solve the equation to determine the value of the unknown number.
    a. 45 is some multiple of 15. Determine the value of the multiple.
    b. $1,200 is 1.5 times as large as some amount of money. What is the amount of money?
    c. 200% of some number is 38.2. What is the number?
    d. 5 is 12 more than ¾ of some number. What is the number?
    e. 10 is 5 times as large as the value that is 4 less than the value of x. What is the value of x?
    f. 7 is 3.5 times as large as some number. What is the number?

7. Evaluate the following expressions:
    a. $-2 + 5 - 12$ b. $6 - (-4) + 1$ c. $2(-3)(-1) - (-6)$ d. $\dfrac{-5 + (9 - 5(4)) + 3}{5(-2) - 1}$

8. Let $y = 3.5x - 7$.
    a. Find the value(s) of y when $x = 0$.   b. What value(s) of x produce a y-value of 11?
    c. What value(s) of y correspond to an x-value of 3?

---

9. Solve each of the equations for the specified variable.

   a. Given $y = 17x - 6$, solve for x when $y = 45$.

   b. Given $y = \dfrac{6x + 5}{3}$, solve for x when $y = 2$.

   c. Given $z = \dfrac{12x - 4}{3}$, solve for x when $z = 10$.

   d. Given $y = \dfrac{3.2x(6 - 0.5x)}{x}$, solve for x when $y = 1$.

   e. Given $y = \dfrac{-6a + 7 - 2a}{2} - 3 + a$, solve for a when $y = 2$.

10. The given graph relates the total value of world exports (internationally traded goods) in billions of dollars to the number of years since 1950.

    a. Define variables to represent the values of the quantities that vary in value in this situation.

    b. Interpret the meaning of the point (17.5, 1000).

    c. As the number of years since 1950 increases from 10 to 20 years, what is the change in the number of years since 1950? Represent this change on the graph.

    d. As the number of years since 1950 increases from 10 to 20 years, what is the corresponding change in the total value of world exports (in billions of dollars)? Explain how you determined this value and represent this change on the graph.

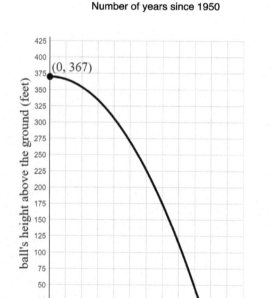

11. A ball is dropped off the roof of a building. The given graph relates the ball's height above the ground (in feet) to the number of seconds that have elapsed since the ball was dropped.

    a. Define variables to represent the values of the quantities with varying values in this situation.

    b. Interpret the meaning of the point (0, 367).

    c. As the number of seconds since the ball was dropped increases from 2 to 4 seconds, how does the height of the ball above the ground change? *Answers may vary slightly since students must estimate values from the graph.*

12. Simplify the following.

    a. $\dfrac{3x^3 + 6x}{x}$

    b. $\sqrt{52x^8 y^3}$

    c. $\dfrac{2x^2 + x - 6}{x + 2}$

13. Simplify the following.

    a. $\dfrac{4(x + 3) + 6x - 12}{x}$

    b. $\sqrt{70x^9 y^{153}}$

    c. $\dfrac{6x^2 - 5x - 21}{(2x + 3)(3x - 7)}$

II. REPRESENTING QUANTITIES AND CHANGES IN QUANTITIES (Text: S1, 3)

14. Matthew started a regular exercise routine and his weight changed from 175 pounds to 153 pounds. What was the change in Matthew's weight?

15. After driving from Tucson to Phoenix, the number of miles on your car's odometer went from 88,312 to 88,428. What was the change in the number of miles on your car's odometer?

16. After spending an hour processing emails, the number of unread emails in your inbox went from 23 to 5. What is the change in the number of unread emails?

17. Let x and z represent the values of two different quantities.
    a. If the value of x decreases from $x = 2$ to $x = -5$, what is the change in the value of x?
    b. If the value of x decreases from $x = 212$ to $x = 32$, what is the change in the value of x?
    c. If the value of z decreases from $z = 2.145$ to $z = 1.234$, what is the change in the value of z?

18. Let a and b represent the values of two different quantities.
    a. If the value of a decreases from $a = 3$ to $a = -4$, what is the change in the value of a?
    b. If the value of a increases from $a = -31$ to $a = 12.2$, what is the change in the value of a?
    c. If the value of b decreases from $b = -1.15$ to $b = -4.21$, what is the change in the value of b?

19. Fill in the given tables showing the appropriate changes in the value of the variable.

s	Δs	y	Δy	p	Δp
−12.43		2.85		3.834	
0.73		14.3		−2.3	
−7.3		−1.05		0	

20. Tom went for a run this morning. Let d represent the number of miles that Tom has run. What is the difference in meaning between $d = 11$ and $\Delta d = 11$?

21. Use the graph to answer the following questions.

    a. Determine Δx and Δy from the point on the left to the point on the right. Illustrate the values of Δx and Δy on the graph.

    b. Determine Δx and Δy from the point on the right to the point on the left. Illustrate the values of Δx and Δy on the graph.

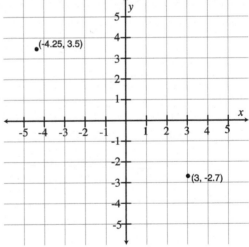

22. The number of calories Monica burns while running increases by 105 calories for every mile that she runs. Let n represent the number of miles that Monica has run. Assume that Monica runs a total distance of 15 miles.
    a. Suppose the value of n increases from $n = 3$ to $n = 7.5$.
        i. What is the change in the value of n?
        ii. What does this change represent in the context of this situation?
        iii. What is the corresponding change in the number of calories Monica burns while running?
    b. How many calories will Monica burn for *any* 4.5-mile change in the number of miles she has run?
    c. For the following changes in the number of miles that Monica has run, determine the corresponding changes in the number of calories that Monica has burned.
        i. $\Delta n = 3.5$ ii. $\Delta n = 0.41$ iii. $\Delta n = 13.7$ iv. $\Delta n = k$ for some constant $k \leq 15$ miles

23. A bucket full of water has a leak. The bucket loses 71 mL of water every 5 minutes. Let *t* represent the number of minutes since the bucket started leaking.
    a. As the number of minutes since the bucket started leaking increases from 13 minutes to 29 minutes, what is the corresponding change in the volume of water in the bucket?
    b. By how much will the volume of water in the bucket change when the number of minutes since the bucket started leaking changes by 1 minute? Explain how you determined your answer.
    c. As the volume of water in the bucket decreases from 60 mL to 23 mL, what is the corresponding change in the number of minutes since the bucket started leaking?
    d. By how much will the number of minutes since the bucket started leaking change when the volume of water in the bucket decreases by 1 mL?

24. A group of Kansas University students were traveling from Lawrence, KS to Denver, CO for a weekend ski trip. On the way they stopped for a late dinner, then continued to Denver driving through the night. They left the restaurant, located 112 miles from Lawrence, at 10:00pm and arrived at Denver, 565 miles from Lawrence, at 5:45 am. Assume the car maintained a constant speed from the time they left the restaurant to the time they arrived in Denver.
    a. Explain what it means to say the car maintained a constant speed from the time it left the restaurant to the time it arrived in Denver. *Make sure to explain the relationship it implies – do not say "the car's speed did not change."*
    b. At what constant speed did the car travel between the restaurant and Denver?
    c. The driver was listening to music to keep awake. Between 2:03 am and 2:55, am the driver listened to his favorite album.
        i. What was the change in the time elapsed while he listened to the album (in minutes)? In hours?
        ii. How far did the car travel while the driver listened to this album?
    d. As the Kansas University students traveled between two towns, they noticed that their trip odometer reading changed from 234.6 miles from Lawrence to 302.4 miles from Lawrence.
        i. What was the change in distance from Lawrence between these two towns?
        ii. How much time elapsed while the car traveled between these two towns?
    e. Sketch a graph of the relationship between the students' distance from Lawrence (in miles) and the number of hours since the students left the restaurant. What does the slope of the graph convey about this situation?

III. CONSTANT RATE OF CHANGE AND LINEAR FUNCTIONS (Text: S2, 3)

25. You are driving on the interstate at a constant speed of 64 miles per hour. How long will it take to drive to the next rest stop that is 16 miles away? Explain the thinking you used to determine your answer.

26. A sports trainer is preparing large containers of a sports drink for a football game. The drink is prepared by mixing powder with water. The instructions call for 13 cups of powder for every 5 gallons of water.
    a. How many cups of powder should be used for 2 gallons of water? Explain your reasoning.
    b. How many cups of powder should be used for 6.3 gallons of water? For 1.8 gallons of water?
    c. How many gallons of water should be used if you have 6 cups of powder? Explain your reasoning.
    d. How many gallons of water should be used if you have 15 cups of powder? 20 cups of powder?
    e. How many cups of powder should be used for 1 gallon of water? How many gallons of water should be used for 1 cup of powder?
    f. If the trainer mixes the sports drink using 8.5 gallons of water and 24 cups of powder, will this mixture taste the same as a mixture made using the instructions? If yes, explain why. If no, explain whether the mixture will taste stronger or weaker than the instructions recommend.

27. When an object is dropped, gravity pulls on the object and causes its speed to increase. The table below shows a certain object's speed at various moments during its fall. Does the object's speed (in feet per second) change at a constant rate with respect to the number of seconds since the object started falling? Explain your reasoning.

Number of seconds since the object started falling	The speed of the object (in feet per second)
0.15	4.83
0.4	12.88
0.52	16.744
0.98	31.556
1.26	40.572

28. The given table of values provides information about an airplane's distance from Sky Harbor International Airport in terms of the number of minutes since the plane took off. Does the airplane's distance from Sky Harbor International Airport change at a constant rate with respect to the number of minutes since the plane took off? Explain your reasoning.

Number of minutes since the airplane took off	The airplane's distance from Sky Harbor International Airport (in miles)
3	16
5	32
9	64
11	92
18	170

29. Given ordered pairs for x and y in the tables, which table(s) contain values that could be from a linear relationship? For any relationship that could be linear, determine the constant rate of change of y with respect to x (that is, the value of m).

Table 1		Table 2		Table 3	
x	y	x	y	x	y
−2	−14	1	3	−5	9.5
3	1	2	7	−2	5
5	7	4	8	−1	3.5
8	16	5	12	3	−2.5
12	28	9	14	10	−13

30. Lisa and Sarah decided to meet at a park bench near both of their homes. Lisa lives 1850 feet due west of the park bench and Sarah lives 1430 feet due east of the park bench. Sarah left her house at 7:00 p.m. and traveled at a constant speed of 315 feet per minute toward the bench. Lisa left her house at the same time traveling a constant speed of 325 feet per minute toward the bench.
   a. Illustrate this situation with a drawing, labeling each quantity that varies in value and each quantity with a fixed value.
   b. Define a formula to represent Lisa's distance (in feet) from the park bench in terms of the number of minutes that have passed since 7:00 pm. *Define any variables you use.*
   c. Define a formula to represent Sarah's distance (in feet) from the park bench in terms of the number of minutes that have passed since 7:00 pm. *Define any variables you use.*
   d. Who will reach the park bench first? Explain your reasoning.
   e. Let's consider how the distance *between* Lisa and Sarah changes as the number of minutes since 7:00 p.m. increases.
      i. Explain how the distance (in feet) between Lisa and Sarah changes as the number of minutes since 7:00 p.m. increases.
      ii. Define a formula that represents the distance between Lisa and Sarah in terms of the number of minutes since 7:00 p.m. *Define any new variables you use.*

31. John and Susan leave a neighborhood restaurant after having dinner. They each walk to their respective homes. John's home is 3120 feet due north of the restaurant and Susan's home is 2018 feet due south of the restaurant. John leaves at 7:28 p.m., traveling at a constant speed of 334 feet per minute, and Susan leaves at 7:30 p.m., traveling at a constant speed of 219 feet per minute.
   a. Illustrate this situation with a drawing, labeling each quantity that varies in value and each quantity with a fixed value.
   b. Define a formula to represent John's distance from the restaurant in terms of the number of minutes that have elapsed since 7:30 pm. *Define any variables you use. Be careful – note the times that each person left the restaurant.*

   c. Define a formula to represent Susan's distance from the restaurant in terms of the number of minutes that have passed since 7:30 pm. *Define any variables you use. Be careful – note the times that each person left the restaurant.*

   d. Will John or Susan arrive home first? Explain your reasoning.

   e. Let's consider how the distance *between* John and Susan changes as the number of minutes that have passed since 7:30 p.m. increases.

      i. Explain how the distance (in feet) between John and Susan changes as the number of minutes that have passed since 7:30 p.m. increases.

      ii. Define a formula that represents the distance (in feet) between John and Susan in terms of the number of minutes that have passed since 7:30pm. *Define any new variables you use.*

For Exercises #32-33: Let r represent the value of one quantity and let p represent the value of another quantity.

32. Let the constant rate of change of r with respect to p have a value of 2.
   a. What does this mean for any change in p?
   b. If p changes by 6, by how much does r change?
   c. If p changes by –3.1, by how much does r change?

33. Let the constant rate of change of r with respect to p have a value of –1.3.
   a. What does this mean for any change in p?
   b. If p changes by 2, by how much does r change?
   c. If p changes by –6.2, by how much does r change?

34. The constant rate of change of y with respect to x has a value of 1.75, and you are given that $y = 12$ when $x = 7$.
   a. What is the value of Δy when $\Delta x = 5$?
   b. What is the value of y when $x = 13$?
   c. What is the value of y when $x = -2$?

35. The constant rate of change of y with respect to x has a value of –3.2, and you are given that $y = -6$ when $x = -1$.
   a. What is the value of Δy when $\Delta x = -7$?
   b. What is the value of y when $x = 2.5$?
   c. What is the value of y when $x = -9.5$?

36. The formula $a = 10 - 1.5t$ represents the remaining length (in inches) of a burning candle, a, in terms of the number of hours that the candle has been burning, t.
   a. What does 10 represent in this context?
   b. What does –1.5 represent in this context?
   c. What does $-1.5t$ represent in this context?
   d. Explain what the point $(t, a) = (2, 7)$ conveys in this context.
   e. What is the value of a when $t = 4.2$? Explain what this value represents in this context.

37. Determine a formula that defines each linear function described in parts (a) through (c).
   a. The constant rate of change of y with respect to x has a value of 2/3 and the vertical intercept is $(0, 5)$.
   b. The constant rate of change of y with respect to x has a value of 5 and the graph crosses the vertical axis at $(0, -2)$.
   c. The constant rate of change of y with respect to x has a value of –6/7 and the vertical intercept is $\left(0, -\frac{1}{10}\right)$.

38. The given graph shows two ordered pairs for a linear function.
    a. Determine the constant rate of change of *y* with respect to *x*.
    b. Using the point $(-5, -0.25)$ as a reference point, what is the change in *x* from this point to $x = 0$? Represent this change on the graph.
    c. What is the change in *y* that corresponds with the change in *x* found in part (b)? Represent this change on the graph.
    d. What is the value of *y* when $x = 0$? Plot the corresponding point on the graph.
    e. Write the formula to represent the value of *y* in terms of *x*.

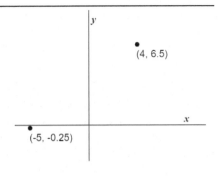

39. The given graph shows two ordered pairs for a linear function.
    a. Determine the constant rate of change of *y* with respect to *x*.
    b. Using the point $(3.6, -7.1)$ as a reference point, what is the change in *x* from this point to $x = 0$? Represent this change on the graph.
    c. What is the change in *y* that corresponds with the change in *x* found in part (b)? Represent this change on the graph.
    d. What is the value of *y* when $x = 0$? Plot the corresponding point on the graph.
    e. Write the formula to represent the value of *y* in terms of *x*.

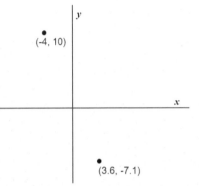

40. Write the formula that defines the linear relationship given in each graph.
    a.   b.

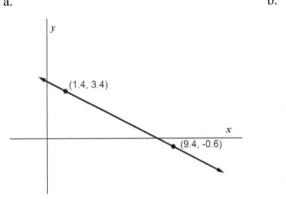
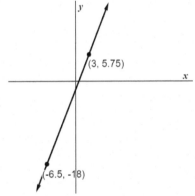

41. The graph of a linear function passes through the points $(2, 9)$ and $(7, -11)$.
    a. What is the constant rate of change of *y* with respect to *x* (slope) for the function?
    b. From the point $(2, 9)$, by how much must *x* change to reach a value of $x = 0$?
    c. What is the corresponding change in the value of *y* for the change in *x* you found in part (b)?
    d. What is the value of *y* when $x = 0$?
    e. Write a formula that represents the value of *y* in terms of the value of *x*.

42. The graph of a linear function passes through the points $(-7, -15)$ and $(5, -7)$.
    a. What is the constant rate of change of *y* with respect to *x* (slope) for the function?
    b. From the point $(-7, -15)$, by how much must *x* change to reach a value of $x = 0$?
    c. What is the corresponding change in the value of *y* for the change in *x* you found in part (b)?
    d. What is the value of *y* when $x = 0$?
    e. Write a formula that represents the value of *y* in terms of the value of *x*.

43. Simplify each of the following expressions.
    a. $2x + 7 - 3x - 2$ b. $\frac{3}{7}x - (-1 + \frac{2}{3}x)$ c. $3x - (-7x) + 4 - 2.2$ d. $2(x - 4) + 3x - (\frac{9}{8}x - 7)$

IV. CONSTANT RATE OF CHANGE & PROPORTIONALITY (Text: S3)

44. A grocery store purchases 150 Honeycrisp apples from a local farmer for $55.50. *Assume there is a proportional relationship between the number of Honeycrisp apples and the total cost of purchasing the apples.*
    a. How many Honeycrisp apples can the store purchase for $250? Explain your reasoning.
    b. How many apples can be purchased for $25? For $100?
    c. How much will it cost to purchase 82 apples? Explain your reasoning.
    d. How much will it cost to purchase 225 apples? 400 apples?
    e. Suppose a different farmer offers to sell the store 200 apples for $70. Is the cost charged by this farmer the same, a better deal, or a worse deal for the store?

45. A photographer has an original photo that is 6 inches high and 10 inches wide and wants to make different-sized copies of the photo so that the new photos are not distorted. (*Assume this constraint for all questions in this exercise.*)

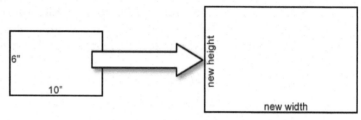

   a. The photographer wants to enlarge the original photo so that the new photo has a width that is 25 inches. What will the height of the new photo need to be so that the image is not distorted? Explain your reasoning.
   b. What ratio must remain constant to assure that the photo is not distorted as it is resized?
   c. Create a table that gives the heights of resized photos corresponding to the new desired photo widths of 1 inch, 2 inches, 5 inches, and 25 inches.
   d. Construct a graph representing the photo's height (in inches) in terms of the photo's width (in inches) for non-distorted photos. Be sure to label your axes with the appropriate quantities and units.
   e. Pick a point on your graph and explain what it represents in this context.
   f. Define variables to represent values of the resized photo's height, h, and width, w. Then write a formula to represent the height of a resized photo in terms of the width of a resized photo for non-distorted photos.
   g. Fill in the blank: The height of a resized photo is always _____ times as large as the width of a resized photo, given that the original photo is 10" wide and 6" high.
   h. Fill in the blank: The ratio of the width to the height of a resized photo is always _____.
   i. Illustrate the following on the graph you constructed in part (d).
      i. The point that represents a 5-inch wide and 3-inch high photo on your graph.
      ii. A line segment that represents an increase of 2.5 inches in the photo's width (from a 10-inch wide photo), and a line segment that represents the corresponding change in the photo's height.

   j. Will any change of 2.5 inches in width always correspond to a change of 1.5 inches in height? Why or why not? How can you see this on the graph?
   k. If the change of the width of the photo is three times as large as a previous change in width, will the corresponding change in height also be three times as large as the previous change in height? Explain.
   l. Fill in the blank: For any change in the resized photo's width, Δw, the change in the resized photo's height, Δh, is always _____ times as large. How can you see this on your graph?

46. A photo printing shop can print photos in any size that a customer requests. Suppose you take a photo with a digital camera using a 4:3 ratio. (This means that the photo's width is $\frac{4}{3}$ times as large as the photo's height, or that the photo's height is $\frac{3}{4}$ times as large as the photo's width). You would like to get your photo printed, but you do not want the image to be distorted or cropped.
    a. If you want to print a photo 12 inches wide, what will its height be if it's not distorted? Explain your thinking.
    b. Construct a table that relates the photo's height, *h*, and the photo's width, *w* (both measured in inches) for photo widths *w* of 4 inches, 7 inches, 10 inches, and 14 inches.
    c. Complete the following statement: As the width of the photo changes, the height of the photo changes so that …
    d. Write a formula that represents the photo's height in terms of its width such that the photo will not be distorted from the original.
    e. Draw a graph representing the photo's height in terms of its width for undistorted photos with a width up to 14 inches. Be sure to label your axes with the appropriate quantities and units.
    f. Explain what each point on your graph represents.
    g. What is the constant rate of change for the relationship you graphed? What does this tell us?

47. The price of ground beef is $4.50 per pound.
    a. Define an ***expression*** to represent the cost (in dollars) of purchasing *n* lbs. of ground beef.
    b. Define a ***formula*** to represent the cost (in dollars), *c*, of purchasing *n* pounds of ground beef.
    c. Construct a graph to represent the cost (in dollars), *c*, of purchasing *n* pounds of ground beef. (*Be sure to label your axes.*)
    d. Explain why the number of pounds of ground beef purchased, *n*, is proportional to the cost (in dollars), *c*, of this purchase.
    e. If you have 2 pounds of ground beef on the scale, and then add another 1.5 pounds of ground beef to the scale, what is the change in the cost of the purchase, Δc?
    f. Will any increase of 1.5 pounds of ground beef always result in the same increase in cost? Justify your answer.
    g. T or F: If *n* and *c* are proportional, then *n* and *c* are changing together at a constant rate of change. Justify your answer or provide a counter example.

48. Determine if the quantities described in each situation are proportional. Justify your response. (*Hint: Think about how the quantities' values are related as they change together.*)
    a. A home decorating consultant charges clients $45 for each hour of consultation. Determine if the *total cost of the consultation* and the *number of hours of consultation* are proportionally related.
    b. A cell phone plan costs $25 per month plus $0.10 per megabyte of data used. Determine if the *monthly cell phone charge* and the *number of megabytes of data used* are proportionally related.
    c. The recommended amount of coffee grounds needed for each cup of water added to a coffee machine is ¾ of a tablespoon. Determine if the *recommended number of tablespoons of coffee grounds* and the *number of cups of water added to the coffee machine* are proportionally related.

49. Consider the given graphs that express two different relationships between values of Quantity A and values of Quantity B.
    a. Are Quantity A and Quantity B proportionally related in Graph 1? Explain.
    b. Are Quantity A and Quantity B proportionally related in Graph 2? Explain.

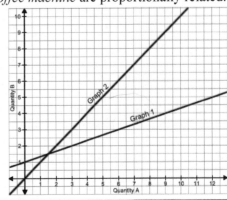

50. Use the following tables of values to answer parts (a) and (b).
    a. Which table(s) give values of quantities that could be related proportionally? Explain your reasoning.
    b. For each table that defines a proportional relationship between two quantities, determine a formula to relate the values of the two quantities.

Table 1		Table 2		Table 3	
x	*y*	*j*	*p*	*b*	*c*
2	12.8	−1	−3.7	−3.1	3.41
4	25.6	2.5	9.25	0.5	−0.55
7	45.8	6.2	22.94	7	−7.7

x	*y*
1	
3	4
7	
12	
18	

51. Complete the following table so that *x* and *y* are proportionally related. Explain how you determined your values of *y*.

52. The given table represents the average weight of 15-year-old males for specific heights.
    a. Are the heights and average weights of 15-year-old males proportionally related? Explain your reasoning.
    b. Is the *change in height* proportionally related to the *change in average weight* for a 15-year-old male? That is, are the quantities related by a constant rate of change? Explain your reasoning.

Height, in inches	Average weight, in pounds
60	109
61	112.7
63	120
66	130.9

53. The post office charge for shipping a package is based on the package's weight. The following table gives the cost of shipping a package for various weights.
    a. If the rate of change of the *total shipping amount charged* (in dollars) with respect to the *weight of the package* (in ounces) is constant, what does this imply about how the two quantities change together?
    b. Is the rate of change of the *total shipping amount charged* (in dollars) with respect to the *weight of the package* (in ounces) constant? If so, determine this constant rate of change.
    c. Are the quantities *weight of the package in ounces* and *total amount charged by the post office* proportionally related? Explain your reasoning.

Weight of the package, in ounces	Total amount charged, in dollars
32	12.80
80	32.00
240	96.00
432	172.80

54. a. Between 2010 and 2020, the Burger Company's profit increased by $35,000 per year. In 2000, the Burger Company's profit was $520,000.
    i. Is *the change in the number of years since 2010* proportional to *the change in the Burger Company's profit*? Explain your reasoning.
    ii. Is *the number of years since 2010* proportional to *the Burger Company's profit*? Explain your reasoning.
    b. You are planning a trip to Las Vegas and need to rent a car. After contacting different companies, you choose to go with the company that charges a $55 rental fee and $0.65 per mile that the car is driven.
    i. Is *the change in the number of miles driven* proportional to *the change in the total cost of renting the car*? Explain your reasoning.
    ii. Is *the number of miles driven* proportional to *the total cost of renting the car*? Explain your reasoning.
    c. When baking chocolate chip cookies, you need 3 cups of flour per cup of sugar.
    i. Is *the change in the number cups of flour* proportional to *the change in the number of cups of sugar* when modifying the recipe for larger batches? Explain your reasoning.
    ii. Is *the number of cups of flour* proportional to *the number of cups of sugar* when modifying the recipe for larger batches? Explain your reasoning.

55. When a bathtub made of cast iron and porcelain contains 60 gallons of water, the total weight of the tub and water is approximately 875.7 pounds. You pull the plug, and the water begins to drain. (*Note that water weighs 8.345 pounds per gallon.*)

Number of gallons of water remaining in the tub	Total weight of the tub and water (in pounds)
59	
40	
30	
20	
10	
8.5	

   a. Describe the quantities in this situation. Which of these quantities have fixed values and which quantities vary in value?

   b. Suppose that some water has drained from the tub and 47 gallons of water remain.

      i. What was the change in the number of gallons of water from when the tub contained 60 gallons of water?

      ii. What is the corresponding change in the total weight of the tub and water?

      iii. What is the weight of the tub and water together when the tub contains 47 gallons of water?

   c. Complete the given table of values.

   d. Suppose you and a friend can each lift about 150 pounds. Once empty, could you and your friend pick up and carry the bathtub out of the bathroom? Explain your reasoning.

56. For each given table,

   a. Determine if the ordered pairs appear to be from a proportional relationship.

   b. Determine if the *changes* in the variables' values appear to be proportional.

Table 1		Table 2		Table 3	
x	*y*	*a*	*b*	*r*	*s*
2	1.42	0.7	1.3	2.4	0.76
−1.2	−0.852	−2	5.62	0.3	3.595
5.1	3.621	5.4	−2.885	−1.5	6.025

   c. Determine if the relationship appears to be linear. If the table appears to represent a linear relationship, write a formula to represent the relationship between the variables' values.

V. EXPLORING AVERAGE SPEED (Text: S4)

57. When running a marathon, you heard the timer call out "12 minutes" as you passed mile-marker 2.

   a. As you passed mile marker 5, you heard the timer call out "33 minutes." What was your average speed from mile 2 to mile 5?

   b. Assume that you continued running at the same constant speed as computed in part (a). How much distance did you cover as your time spent running increased from 35 minutes after the start of the race to 40 minutes after the start of the race?

   c. You passed mile marker 5 after running for 33 minutes. If you completed the race (26.2 miles) in a total of 175 minutes, what was your average speed from the moment you passed mile marker 5 to the end of the race?

   d. What is the meaning of the average speed you calculated in part (c)?

58. When running a road race, you heard the timer call out "8 minutes" as you passed the first mile marker in the race.

   a. As you passed mile marker 6, you heard the timer call out "52 minutes." What was your average speed from mile marker 1 to mile-marker 6?

   b. After mile marker 6, you ran at a constant rate of 10 minutes per mile between mile marker 6 and mile-marker 10.

      i. Was your speed from mile marker 6 to mile-marker 10 faster or slower than your average speed from mile marker 1 to mile marker 6?

      ii. How many minutes did it take you to travel from mile marker 6 to mile marker 9?

   c. If you continued running at the same constant speed after passing mile marker 10, what is the total distance you had traveled from the starting line 130 minutes after starting the race?

59. The given graph represents two relationships:
(i) Kevin's distance traveled since passing mile marker 225 while cycling (in miles) in terms of the number of minutes since passing mile marker 225 and (ii) Carrie's distance traveled since passing mile marker 225 while cycling (in miles) in terms of the number of minutes since passing mile marker 225. *Assume the two cyclists are riding on the same road and in the same direction.*

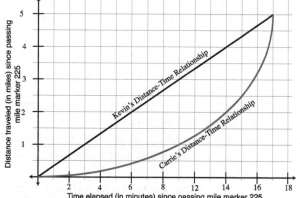

a. What happens 17 minutes after Carrie passed mile marker 225?
b. How do the distances traveled and times elapsed compare for Carrie and Kevin as they traveled from mile marker 225 to mile marker 230?
c. How do Carrie's and Kevin's speeds compare as they travel from mile marker 225 to mile marker 230?
d. How do Carrie's and Kevin's average speeds compare over the time interval as they traveled from mile marker 225 to mile marker 230?
e. If Kevin continued cycling at the same constant speed, how far past mile marker 225 would he be 28 minutes after passing it?

60. Marcos traveled in his car from Phoenix to Flagstaff, a distance of 155 miles.
a. Determine the amount of time required for Marcos to travel from Phoenix to Flagstaff if his average speed for the trip was 68 miles per hour.
b. Construct a possible distance-time graph of Marcos's trip from Phoenix to Flagstaff. Be sure to label your axes.
c. On the same axes, construct a graph that represents the distance-time relationship for a car traveling along the same road, in the same direction, leaving Phoenix at the same time, and driving at a constant speed of 68 miles per hour.
d. What is true about the graphs you drew in parts (b) and (c)?

61. The distance (in feet) from Silvia's house to her current position, d, is modeled by the formula $d = t^2 + 3t + 1$, where t represents the number of seconds since Silvia started walking.
a. What was Silvia's change in distance as the time since Silvia started walking increased from 2 to 7 seconds?
b. Find Silvia's average speed for the period from $t = 2$ to $t = 7$ seconds.
c.   i. Construct a graph that gives Silvia's distance from her house (in feet) in terms of the number of seconds since she started walking. Be sure to label your axes.
  ii. Illustrate on the graph in part (i) the change in the number of seconds Silvia walked and the corresponding change in Silvia's distance from her house over the period from $t = 4$ to $t = 5$.
  iii. Illustrate on the graph in part (i) the change in the number of seconds Silvia walked and the corresponding change in Silvia's distance from her house over the period from $t = 3$ to $t = 7$.
  iv. Determine Sylvia's average speed over each of the two intervals in parts (ii) and (iii).

62. Bob's distance north of Mrs. Bess's restaurant (in feet), d, is given by the formula $d = 2t^2 - 7$, where t represents the number of seconds since Bob began driving.
a. Determine the value of d when $t = 1$. What does a negative value for d represent in the context of this problem?
b. Find the average speed of Bob's car for the period from $t = 3$ to $t = 5$.
c. As the number of seconds since Bob began driving increased from 1.5 to 2 seconds, by how much did Bob's distance north of Mrs. Bess's restaurant change?

d. As the number of seconds since Bob began driving increased from 2 to 2.25 seconds, by how much did Bob's distance north of Mrs. Bess's restaurant change?

e. i. How much time did it take Bob to travel from 20 to 30 feet north of the restaurant?

ii. How much time did it take Bob to travel from 30 to 40 feet north of the restaurant?

iii. How much time did it take Bob to travel from 40 to 50 feet north of the restaurant?

63. The given graph represents the speeds of two cars (car A and car B) in terms of the elapsed time in seconds since leaving a rest stop. Car A travels at a constant speed of 65 miles per hour. As car A passes the rest stop, car B pulls out beside car A and they both continue traveling down the highway.

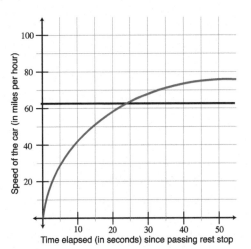

a. Which graph represents car A's speed and which graph represents car B's speed? Explain.

b. Which car is further down the road 20 seconds after being at the rest stop? Explain.

c. Explain the meaning of the intersection point.

d. What is the relationship between the positions of car A and car B at the graph's intersection point?

For Exercises #64–68, let *d* be the distance of a car (in feet) from mile marker 420 on a country road, and let *t* be the time elapsed (in seconds) since the car passed mile marker 420. The given formulas represent various ways these quantities might be related. For each formula, do the following:

i. Determine the car's average speed over the specified interval.

ii. Explain the meaning of the average speed you calculated in part (i).

64. $d = t^2$ from $t = 5$ to $t = 30$.

65. $d = -3(-19t - 1)$ from $t = 3$ to $t = 9$.

66. $d = 5(12t + 1) + 3t$ from $t = 0.5$ to $t = 3.75$.

67. $d = \dfrac{10t(t + 5) - 14}{2}$ from $t = 0$ to $t = 5$.

68. $d = (2t + 7)(3t - 2)$ from $t = 2$ to $t = 2.75$.

VI. THE DISTANCE FORMULA (Text: S5)

69. Use the distance formula to find the distance between each pair of given points. *Your answer should include six distances.*

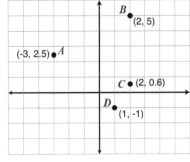

70. A circle is centered at the origin $(0, 0)$ and has a radius of length 4.

a. What is the equation for the graph of this circle?

b. For which points on the circle is the *x*-coordinate equal to 1?

c. For which points on the circle is the *y*-coordinate equal to 3?

71. A circle is centered at the origin $(-2, 4)$ and has a radius of length 6.

a. What is the equation for the graph of this circle?

b. For which points on the circle is the *x*-coordinate equal to 0?

c. For which points on the circle is the *y*-coordinate equal to 4?

72. The center of a circle is located at the point $(1, 1)$. The point $(-5, 2)$ is located on the circle.

a. What is the circle's radius length?   b. What is the equation for the graph of this circle?

73. The center of a circle is located at the point (–5, 0). The point (–9, –3) is located on the circle.
    a. What is the circle's radius length?   b. What is the equation for the graph of this circle?

VII. ABSOLUTE VALUE (Text: S6)

74. A contractor is digging a hole that needs to be 36" deep. He digs a hole that is no more than 1.5 inches different from the necessary depth.
    a. Indicate on a number line the possible depths for the hole he dug.
    b. Use algebraic symbols to represent values of x (hole depths, in inches) that are within 1.5 inches of the desired depth.
    c. Write an expression to represent the difference between the depth of hole the contractor dug in inches, x, and the desired depth.
    d. Use algebraic symbols to represent that the difference between the depth of the hole dug and the desired depth must be within 1.5 inches.

75. The mass of a bag of cookies is 541 grams according to the label. The actual mass may vary from the label value by no more than 3 grams for the bag of cookies to be sold.
    a. Represent the possible weights of a bag of cookies acceptable for sale on a number line.
    b. Use algebraic symbols to represent values of w (the mass of a bag of cookies in grams) that correspond to bags that can be sold.
    c. Write an expression to represent how much the mass of a bag of cookies (in grams), w, differs from the mass on the label.
    d. Use algebraic symbols to represent that the difference between the mass of a bag of cookies and the mass listed on the package must be no more than 3 grams.

76. Represent all numbers whose distance from 3 is less than 4 on a number line.

77. Represent all numbers whose distance from –5 is less than 2 on a number line.

78. What do the solutions to the inequality $-4 < x - 3 < 4$ represent?

79. What do the solutions to the inequality $-2 < x - (-5) < 2$ represent?

80. Determine all values of x that satisfy the given constraint.
    a. All numbers x whose distance from 7 on the number line is less than 4 units.
    b. All numbers x whose distance from –2 on the number line is no more than 3.4 units.
    c. All numbers x whose distance from –5 on the number line is at least 5.2 units.

81. Illustrate the solutions to each of the given absolute value equations on a number line.
    a. $|x| = 3.5$ b. $|x| < 3.5$ c. $|x| > 3.5$

82. Represent the solution set of each of the following absolute value inequalities by:
    i. describing the solution in words,
    ii. illustrating the solutions on a number line, and
    iii. writing an inequality (with no absolute values).
    a. $|x - 2| < 4$ b. $|x + 7.2| < 3$ c. $|x - 14| > 5$ d. $|x + 7.3| > 2.7$

83. Solve each of the following equations for x. Check your answers and show your work.
    a. $|x| = 7$ b. $|x - 5| = 3$ c. $|4x - 7| = 15$ d. $|11x + 22| = -44$ e. $|2x - 2| = x + 3$

This investigation contains review and practice with important skills and procedures you may need in this module and future modules. Your instructor may assign this investigation as an introduction to the module or may ask you to complete select exercises "just in time" to help you when needed. Alternatively, you can complete these exercises on your own to help review important skills.

Evaluating and Simplifying Expressions and Solving Equations
Use this section prior to the module or with/after Investigation 1.

In Exercises #1-3, evaluate each expression given that $a = 4$, $b = 11$, $c = 5$, and $d = -4$.

1. $\dfrac{\sqrt{c-a}+b}{2}$

2. $\dfrac{(a+1)-a}{b^2-b}$

3. $\dfrac{d\sqrt{d+6}}{d+6}$

In Exercises #4-7, use the given information to write an equation and then solve it.

4. solve $y = 2x - 19$ for x if $y = 5$

5. solve $y = \sqrt[3]{x}$ for x if $y = -2$

6. solve $n = -\dfrac{1}{3}(p-11)+19$ for p if $n = 7$

7. solve $\dfrac{w+2}{w+6} = q$ for w if $q = -1$

In Exercises #8-11, solve each equation.

8. $\dfrac{2}{3}x + \dfrac{1}{2} = \dfrac{17}{2}$

9. $4 = \dfrac{1}{5}(3n-7)$

10. $x + 3 = \dfrac{x-3}{4}$

11. $\dfrac{3}{4}x + \dfrac{5}{6} = 5x - \dfrac{125}{3}$

In Exercises #12-16, write an equation based on each description and then solve the equation.

12. The sum of 3 times some number and 12 is 42. What is the number?

13. The difference between some number and 14.3 is 2.1. What is the number?

14. $1,200 is 1.5 times as large as some number. What is the number?

15. 200% of some number is 38.2. What is the number?

16. Three-fourths of some number increased by 12 is equal to 5 times the number. What is the number?

Evaluating and Simplifying Expressions and Solving Equations (with Function Notation)
Use this section with/after Investigation 2.

17. Given that $g(x) = \dfrac{3x}{x-7} + 2$, evaluate each of the following.

   a. $g(8)$ b. $g(10)$ c. $g(-1.5)$ d. $-\dfrac{g(4)}{4}$

In Exercises #18-19, use the given information to write an equation and then solve it.

18. if $h(x) = 3(7-x) + 4$, find the value of x such that $h(x) = 31$.

19. if $j(x) = \dfrac{1}{x+2}$, find the value of x such that $j(x) = -0.5$.

Modeling and Additional Content
Use this section prior to the module or with/after any of the first three investigations.

20. We define the following variables to represent dimensions of a circle that can change.
   r = circle's radius length measured in feet
   d = circle's diameter length measured in feet
   C = circle's circumference length measured in feet
   A = circle's area measured in square feet

   a. Write the formula to represent a circle's diameter length in terms of its radius length. Use the formula to determine a circle's diameter length if its radius is 4.721 feet long.

   b. Write the formula to represent a circle's circumference length in terms of its diameter length. Use the formula to determine a circle's circumference length if its diameter is 6.48 feet long.

---

c. Write the formula to represent a circle's area in terms of its diameter length. Use the formula to determine the area of a circle if its diameter is 4.09 feet long.

d. Write the formula to represent a circle's circumference length in terms of its radius length. Use the formula to determine a circle's circumference length if its radius is 3.5 feet long.

e. Define a formula to represent a circle's circumference length in terms of its area. Use the formula to determine a circle's circumference length if its area is 42.7 square feet.

21. a. The given graph represents how values of x and y are related. Determine the value of x when $y = 8$.

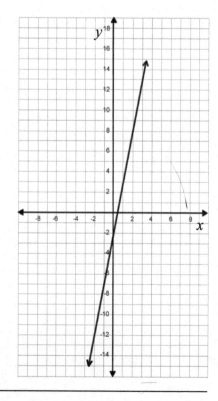

b. Given that $y = 5x - 2$ is the formula of the graphed function, solve the equation $8 = 5x - 2$ for x algebraically. Illustrate how to use the graph to solve this equation.

c. Use the graph to solve the equation $3 = 5x - 2$. Illustrate how you used the graph to solve this equation.

d. Use the graph to solve the equation $0 = 5x - 2$. Illustrate how you used the graph to solve this equation.

*1. Consider what is involved in building a box (without a top) from an 8.5" by 11" sheet of paper by cutting squares from each corner and folding up the sides.

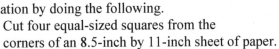

8.5"

11"   the cutout

a. To understand how the quantities in the situation are related, it is important to first model the situation by doing the following.
   i. Cut four equal-sized squares from the corners of an 8.5-inch by 11-inch sheet of paper.
   ii. Fold up the sides and tape them together at the edges.

b. Do the cutouts have to be square? Explain.

   Yes

c. Describe each quantity with a varying value. Then describe each quantity with a fixed value.

   Varying   Fixed

   Height   Original length + width

   Weight

d. Describe how the configuration of the box changes as the length of the side of the square cutout increases? What is the largest value of x, the square cutout side length (in inches), that makes sense?

   Box gets higher/taller as cutout increases

   Largest value is 4.25

e. Determine the values of each of the following for your box.

   Box's height: _3 in_   Length of the box's base: _5 in_

   Width of the box's base: _2.5 in_   Volume of the box: _37.5 in³_

f. Describe how the volume of the box varies as the length of the side of the cutout varies from 0 to 4.25 inches.

   Volume

g. Based on your responses to previous parts, draw a rough sketch of a graph of the volume of the box (in cubic inches) in terms of the length of the side of the square cutout (in inches). *Be sure to scale and label your axes.*

h. Use a ruler to measure the length of the side of the square cutout (in inches) of your box.
Answer: _____

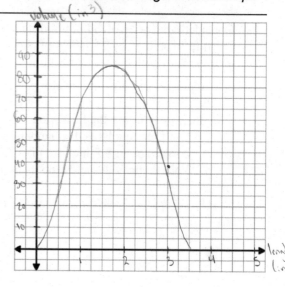

*2. Let x represent the length of the side of the square cutout in inches. Let w represent the width of the box's base in inches. Let l represent the length of the box's base in inches. Let V represent the volume of the box in cubic inches.

a. Complete the table of values.

x	w	l	V
0			
	6.5		
2.5			
4.25			
		2	

b. Explain how the length of the base of the box is related to the length of the side of the square cutout and how the width of the base of the box is related to the length of the side of the square cutout.

c. Construct an expression to represent the width of the box in terms of the length of the side of the square cutout, x.

$$w - 2x$$

d. Construct an expression to represent the length of the box in terms of the length of the side of the square cutout, x.

$$l - 2x$$

e. Define a formula to determine the volume of the box in terms of the length of the side of the square cutout.

f. Use your formula from part (e) to represent the volume of the box when x, the length of the side of the cutout, is 0.5 inches.

g. Use your formula from part (e) to represent the volume of the box when x, the length of the side of the cutout, is 3 inches.

3. a. Create a graph that represents the volume of the box, V (measured in cubic inches), in terms of the length of the side of the square cutout, x (measured in inches).

    (When determining the window setting on your calculator, consider the possible values of x and the possible values of V that make sense.)

    Construct the graph and label points on the graph that correspond to boxes you (or the class) built. State what two of these points convey about the box.

    Point 1:

    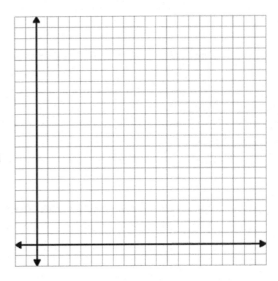

    Point 2:

    b. As x (the length of the side of the cutout) increases from 0.5 inch to 0.75 inch, how does the volume of the box change? Indicate these changes on your graph from part (a).

    c. As x (the length of the side of the cutout) increases from 1 inch to 3 inches, how does the volume of the box change?

d. As x (the length of the side of the cutout) increases from 2.1 inches to 2.7 inches, how does the volume of the box change?

e. Indicate on the graph i) a change of cutout length from 1 inch to 2.5 inches and ii) the corresponding change in the box's volume.

f. Estimate the interval(s) of values for the length of the side of the cutout x for which the volume of the box decreases as x increases.

*4. a. Using your graphing calculator, solve for the possible lengths of the side of the square cutout if the volume of the box is 62.5 cubic inches.

b. Using your graphing calculator, solve for the possible lengths of the side of the square cutout if the volume of the box is 23 cubic inches.

c. Using your graphing calculator, find an approximate maximum value for the box's volume.

d. Using your graphing calculator, solve for the length of the side of the square cutout that produces the maximum volume you found in part (c).

5. a. If the length of the side of the square cutout is 3 inches, what is the box's exterior surface area (in square inches)? *Draw a diagram to help you visualize the box's exterior surface area.*

   b. How does the box's exterior surface area vary with the length of the side of the square cutout?

   c. Define a formula to determine the box's exterior surface area, S (in square inches), in terms of x, the length of the side of the square that is cut from each of the four corners.

   d. What is the exterior surface area of the box with the maximum volume? Explain the thinking you used to determine your answer.

   e. What is the volume of the box that has an exterior surface area of 25 square inches?

In previous investigations we wrote formulas to describe how the value of one quantity is related to the value of another quantity as the values change together. We say that the dependent quantity is a *function* of the independent quantity if every value of the independent quantity produces *exactly one* value of the dependent quantity. In this investigation we will introduce the uses and benefits of ***function notation*** and practice representing formulas that assign each value of the independent quantity to exactly one value of the dependent quantity using function notation.

It is also convention to say that a value of the independent quantity is an ***input*** value to ("put into") the function rule, and a value produced from applying the function rule is an ***output*** value of the function rule.

*1. In the previous investigation we defined the volume of a box formed by cutting squares from an 8.5" by 11" sheet of paper and folding up the sides with the formula:

$$V = x(11 - 2x)(8.5 - 2x)$$

a. For each value of x, determine the corresponding box's volume (in in^3), V.

$x = \underline{1.5}$ inches, $V =$ _____ cubic inches

$x = \underline{3.1}$ inches, $V =$ _____ cubic inches

b. Is it possible to use the formula $V = x(11 - 2x)(8.5 - 2x)$ to represent the value of V when $x = 1.5$ without calculating that value?

c. *Function notation provides a concise way to reference the value of a dependent quantity that is associated with a particular value of an independent quantity <u>without actually computing its values</u>.*

For the volume formula for the box problem, we can write $f(x) = x(11 - 2x)(8.5 - 2x)$, instead of $V = x(11 - 2x)(8.5 - 2x)$.

We read $f(x)$ as "f of x". The symbol f is designated to be the function name, and $f(x)$ represents the values of the function's volume. We can choose a specific value for x and input it into the function to discuss a value of the dependent quantity we are interested in. For example, $f(2.5)$ represents the box's volume when the square's side length x is 2.5 inches.

i. What does $f(1.5)$ represent in the context of the box problem? How do you read $f(1.5)$?

ii. What does $f(3.1) - f(1.5)$ represent? How do you read $f(3.1) - f(1.5)$?

Formulas with Function Notation

Let x represent the length of the side of the square cutout (in inches) and V represent the box's volume (in cubic inches). Call the relationship between these quantities f, where x represents input values and V represents output values. Then:

$$f(x) = x(11 - 2x)(8.5 - 2x) \quad \text{with} \quad f(x) = V$$

Let's clarify the meaning of this notation.

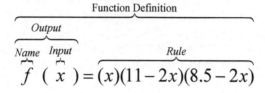
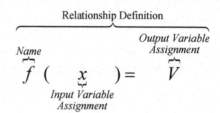

When a function is defined with a formula, we produce a rule that tells us how to match each value in the **domain** (values of the independent quantity) to the corresponding value in the **range** (values of the dependent quantity).

---

*2. Read the above "Formulas with Function Notation" box, then answer the following questions.

  a. Using function notation and the function name f, we define the box's volume, V, in terms of the length of the side of the square cutout, x, as $f(x) = x(11 - 2x)(8.5 - 2x)$ with $V = f(x)$.

   i. What is the independent (input) quantity?   ii. What does $f(x)$ represent?

   iii. What does f represent?   iv. What is the domain of f?

  b. Illustrate each of the following on the given graph of f.

   i. the point $(3.5, f(3.5))$ ii. $f(1.5)$

   iii. the solutions to $f(x) = 40$

  c. Use the function formula $f(x) = x(11 - 2x)(8.5 - 2x)$ to evaluate $f(2)$ and describe what this value represents.

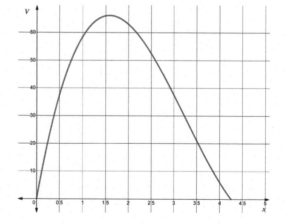

  d. Using function notation, represent the change in the box's volume as the length of the side of the square cutout increases from 2 inches to 3.5 inches. Then represent this change on the graph that is given in part (b).

<u>Function</u>

A **function** consists of three parts:
1. *Domain*: The values the independent quantity may assume.
2. *Range*: The values the dependent quantity may assume.
3. *Rule*: The rule assigns each value of the independent quantity to *exactly one* value of the dependent quantity.

The rule of a function can be expressed using any of the following: i) a worded description; ii) an algebraic expression; iii) a graph; or iv) a table of values. The function rule defines how a function's independent and dependent quantities are related as their values change together.

Terminology: If we want to write a function f to represent a square's perimeter length P (in inches) in terms of its side length s (in inches), we write $f(s) = 4s$ with $P = f(s)$. (*Both P and $f(s)$ represent the value of the square's perimeter length.*)

We say that the function f "defines $f(s)$ in terms of s," meaning the value of the dependent quantity (the perimeter), represented by both $f(s)$ and P, is determined by (or depends on) the value of the independent quantity (the side length), represented by s.

Some advantages of using function notation include:
- We can reference the function relationship using its name, f. This is more concise than having to say or write out $P = 4s$ or a description of the relationship.
- We can reference a function's output value that corresponds to a particular input value without having to actually compute the value. So, $f(3.2)$ represents a square's perimeter length (in inches) with a side length of 3.2 inches even if we don't calculate its value.

---

*3. Read the definition and terminology overview above.
   a. What does $f(5.8)$ represent in the scenario presented?

   b. If we are given that the square's perimeter length cannot exceed 48 inches…
      i. what are the possible values that the square's side length s can take on? (*Your answer describes the **domain** of f.*)

      ii. what are the possible values that the perimeter P can take on? (*Your answer describes the **range** of f.*)

   c. What is the rule that determines how s and $f(s)$ are related and change together?

---

d. T or F: For function f there is exactly one value of $f(s)$ for each value of s. Justify your choice.

The Vertical Intercept of a Function

A function's *vertical intercept* is the ordered pair corresponding to a value of 0 for the independent (input) quantity. For a function f, the vertical intercept is $(0, f(0))$.

For example, if $f(0) = -25$, the vertical intercept of f is $(0, -25)$, and when we graph this point it will fall on the vertical axis.

---

4. a. The function h defined by $h(x) = 39 - 6.5x$ represents the distance (in meters) that a tortoise is ahead of a hare in terms of the number of seconds since the start of a 100-meter race, x. Construct the graph of h. *Be sure to label your axes.*

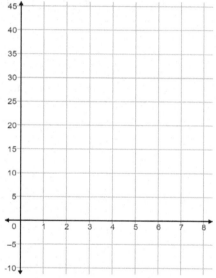

b. Use the graph to approximate the value of $h(4.5)$ and explain what the answer represents.

c. Use the algebraic representation of function h to evaluate $h(4.5)$ and compare this value with the one you determined/estimated from the graph. If the answers differ, explain how this could be the case.

d. Determine the vertical intercept of h and describe what it represents in the context of this problem.

Horizontal Intercepts of a Function

A function f's *horizontal intercept(s)* are the ordered pairs corresponding to a value of 0 for the dependent (output) quantity. In other words, they have the form $(a, 0)$ for some real number a.

For example, if $f(-7) = 0$, a horizontal intercept for f exists at $(-7, 0)$, and when we graph this point it will fall on the horizontal axis.

For example, given a function such as $h(x) = x(x - 7)$, we can determine the **horizontal intercepts** by determining the values of x that make $h(x) = 0$. To find the horizontal intercepts, we solve the equation $0 = x(x - 7)$. Since this equation is true when either factor is 0, the solutions are $x = 0$ and $x = 7$. The values $x = 0$ and $x = 7$ are also called **roots** or **zeros** of the function h, and the horizontal intercepts will occur at $(0, 0)$ and $(7, 0)$.

*5. Continuing the same context from Exercise #4, $h(x) = 39 - 6.5x$ represents the distance (in meters) that a tortoise is ahead of a hare in terms of the number of seconds since the start of a 100-meter race, x.

    a. Solve the equation $h(x) = 0$ for x. What does this solution represent in this context? Label this solution on the graph you created in Exercise #4, part (a).

    b. What is the root of h? What point represents the horizontal intercept of the graph of h?

6. For each of the following functions, determine all horizontal intercepts and explain what they represent. *Try to determine these algebraically first, then verify your answers by graphing the functions on your calculator.*
    a. $h(x) = 4(x - 6) + 2$ b. $b(x) = x(11 - 2x)(8.5 - 2x)$

*7. For this graphical representation of the relationship between x and y...
    a. Determine if y is a function of x. Justify your answer using the definition of a function—that is, each input value of a function is assigned to exactly one output value.

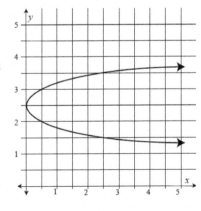

    b. Determine if x is a function of y. Justify your answer.

Recall that the **domain** of a function is the set of all possible values of the independent quantity for which the function is defined. A function's domain can be restricted for many reasons. Some examples include:

i) it is not possible to determine a real-number square root of a negative number;

ii) the denominator of a quotient cannot be 0; or

iii) values for the independent quantity are restricted by the context (for example, the cutout length in the box problem context cannot exceed 4.25).

*8. Without using a graphing calculator, determine each function's domain. Then, with the assistance of a graphing calculator or graphing software, determine each function's range.

a. $s(x) = \dfrac{9}{x}$

b. $g(x) = \dfrac{1}{x^2 - 9}$

c. $h(x) = x^2 + 2x - 5$

d. $k(x) = \dfrac{\sqrt{x-2}}{x-9}$

9. Without using a graphing calculator, determine the domain and range of each of the following functions.

a. $p(x) = \dfrac{x}{9}$

b. $f(x) = \sqrt{x-4}$

c.

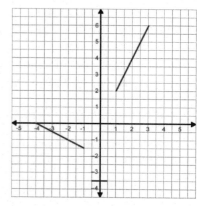

*1. If A is the area of a square measured in square inches and s is the square's side length measured in inches, then the square's side length, s, and its area, A, are related by the formula $A = s^2$. If we call this function g, recall that using function notation we can also write $g(s) = s^2$ where $A = g(s)$.

a. Use function notation to represent (NOT CALCULATE) the areas of two different squares with side lengths 3.5 inches and 26.92 inches.

b. Interpret the meaning of $g(4.9) = 24.01$ in this context.

c. What does it mean to solve $g(s) = 81$ for s? (*Don't solve yet and don't explain __how__ to solve for s, instead explain what the solution __represents__.*)

d. Solve for s when $g(s) = 24.01$. (Recall that $g(s) = s^2$, where the square's area $A = g(s)$.)

e. Explain what each of the following represents.
   i. $g(5) + 23$ ii. $4g(3.4)$

*2. *Continue with the same context from Exercise #1:* If A is the area of a square measured in square inches and s is the square's side length measured in inches, then the square's side length, s, and its area, A, are related by the formula $A = s^2$. If we call this function g, recall that using function notation we can also write $g(s) = s^2$ where $A = g(s)$.

a. If the square's side length s begins at 0 inches and starts to increase at a constant rate of 3 inches per second, where t represents the number of seconds since the square's side length started to increase, what does $3t$ represent?

b. What does $(3t)^2$ represent in this situation? Expand the expression $(3t)^2$ and explain what this new expression represents and how it compares to the expression $(3t)^2$.

*3. The amount of medication (per dose of a specific medicine) needed for a patient varies with the patient's weight. If *g* determines the number of ounces of medication in one dose in terms of the patient's weight, *x* (in pounds), determine what each of the following represents.

a. *g(x)*

function that determines medication needed based on pound of patient's weight

b. *x*

weight of patient (lbs)

c. *g*

name of function

d. *g(145)*

oz of med needed for someone who is 145 lbs

e. *1.5g(85)*

patient needs 1.5 x normal oz of meds

1.5 x oz of medicine for patient weighing 85 lbs

f. *g(110) + 1.5*

oz meds needed for patient weighing 110 lbs plus 1.5 oz of meds

g. *g(195) = 2.5*

patient weighing 195 lbs needs 2.5 oz meds

h. The solution to the equation *g(x) = 3.25*.

weight of the patient who gets 3.25 oz of meds

4. The given table represents the retail price (in dollars) of a brand-new Toyota Camry as a function of the number of years after 2000, *n*.

# of years since 2000, *n*	Price of car (dollars), *f(n)*
0	17,520
5	19,295
7	19,995
10	20,835
13	21,985
20	25,925
24	27,515

a. What is the input quantity? What is the output quantity?

input = # of years since 2000

output = price of car

b. Evaluate each of the following and describe the meaning in this context.

i. *f(7) =* $19,995

7 years after 2000, the car costs $9,995

ii. *f(20) =* $25,925

20 years after 2000, the car costs $25,925

c. Solve for *n* when *f(n) = 19295* and explain what your solution represents.

n = 5 years

d. Evaluate each expression and describe its meaning in this context.

   i. $f(7) - f(5)$

   ii. $f(5) + 1250$

   iii. $\dfrac{f(24)}{f(10)}$

*5. A tortoise and hare are competing in a race around a 400-meter track. The arrogant hare decides to let the tortoise have a 225-meter head start. When the start gun is fired, the hare begins running at a constant speed of 5.5 meters per second and the tortoise begins crawling at constant speed of 2.0 meters per second. Let t represent the number of seconds since the start gun was fired. We are interested in determining who won the race.

   a. Illustrate the situation with a diagram (mark off the length of the race, note the approximate position of the tortoise from the starting line and the position of the hare, illustrate quantities with varying values with dashed arrows, etc.).

   b. Define a function g to represent the tortoise's distance from the finish line in terms of the number of seconds since the start gun was fired, t.

   c. Define a function h to represent the hare's distance from the finish line in terms of the number of seconds since the start gun was fired, t.

d. Evaluate $g(40)$ and say what your answer represents.

e. Solve $g(t) = 80$ for t and say what the solution represents.

f. Solve $h(t) = 0$ and say what the solution represents.

g. Construct a graph of g and h on the given axes.

h. Evaluate $h(20) - g(20)$ and indicate how this quantity is represented on the graph in part (g). What does the quantity represent in the context of this situation?

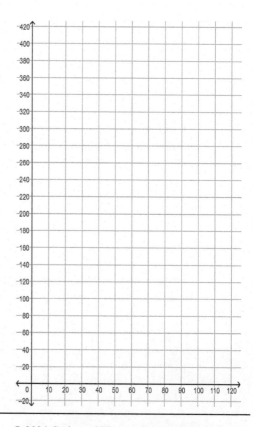

i. Define a new function f where $f(t) = h(t) - g(t)$, and graph this function on a new set of axes. What does function f represent?

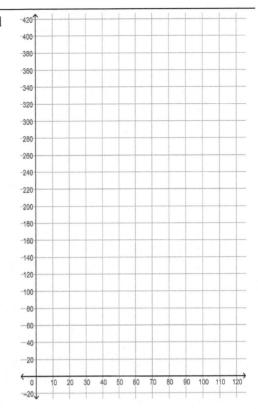

j. What is the slope of the graph of f? What does this value represent in the context of this situation? How does the distance between the tortoise and hare change throughout the race?

k. Solve $g(t) = h(t)$ for t and say what the solution represents.

l. Who won the race? Justify your answer using multiple methods.

6. A hose is used to fill an empty wading pool. The graph shows the volume of the water added to the pool from this hose, V (in gallons), as a function of the elapsed time (in minutes) since the hose was turned on, t.

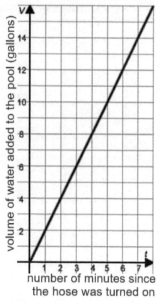

a. Define a function g that determines the volume of water (in gallons) added to the pool as a function of t, the elapsed time (in minutes) since the pool started filling.

b. Evaluate $g(4) - g(1)$ and represent this difference on your graph. Describe what the expression $g(4) - g(1)$ represents in this problem context.

c. What does $g(4) + 100$ represent in this context?

d. After turning on the first hose, a second hose is turned on and is placed in the pool. Given that the function f, defined by $f(t) = g(t - 4)$ with $t \geq 4$, determines the amount of water that has flowed into the pool from the second hose t minutes since the first hose was turned on, complete the following.

    i. At what time was the second hose turned on? What is the domain of f?

    ii. What does the expression $g(t) + f(t)$ represent?

    iii. How does the graph of f compare to the graph of g?

    iv. Evaluate $g(5) + f(5)$ and explain what your answer represents.

*1. Running is a popular form of exercise to burn calories and stay healthy. The number of calories burned while running depends on many factors but averages about 110 calories per mile. Suppose Mila goes for a run, traveling at a constant speed of 0.15 mile per minute, while burning approximately 110 calories per mile.

a. How many calories does Mila burn if she runs for 40 minutes? Explain the two-step process you used to determine your answer.

0.15 mile/min $\cdot$ 40 min $= 6 \times 110$ cal/mile

660 calories in 40 minutes

b. How many calories does Mila burn if she goes for a 34-minute run? Explain the two-step process you used to determine your answer.

0.15 mile/min $\cdot$ 34 min $= 5.1$ miles

5.1×110 cal/min $= 561$ calories in 34 min

c. Describe a two-step process for determining the number of calories Mila burned when running for t minutes at a speed of 0.15 mile per minute.

$f(t) = (0.15t) \cdot 110$ OR $f(t) = 16.5t$

d. Define a function f that represents the distance (in miles) Mila has run, d, in terms of the number of minutes Mila has run, t.

$f(t) = 0.15t$ $\rightarrow f(t) = d$

e. Define a function g that represents the number of calories Mila burned, c, in terms of the number of miles she ran, d.

$g(d) = 110d$ $\rightarrow g(d) = c$

f.   T or F: The output quantity of function *f* is the input quantity for the function *g*. Justify your choice.

g.   We have described a two-step process for determining the number of calories that Mila burns when running for *t* minutes. Label the input and output quantities for each function process in the given illustration.

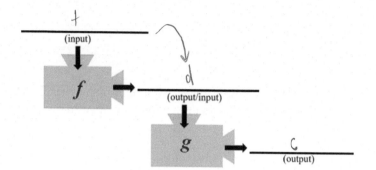

h.   Define a function *h* that represents the number of calories Mila burns, *c*, in terms of the number of minutes Mila has run, *t*.

$$h(t) = 16.5t \qquad (110 \cdot 0.15) \qquad\qquad h(t) = c$$

$$h(t) = 110(0.15t) \quad (\text{both operations})$$

$$h(t) = 110\,f(t)$$

$$h(t) = g\big(f(t)\big) - \text{function composition would ONLY work if one output becomes the next input}$$

There are two key takeaways from Exercise #1.

(I)   The expressions we write should be meaningful and related to our understanding of the quantities. For example, note the meaning of each of the following expressions and how new meanings develop as we modify them.

- *t* represents the number of minutes Mila has been running.
- $0.15t$ represents the distance (in miles) Mila runs in *t* minutes.
- $110(0.15t)$ represents the number of calories Mila burns in *t* minutes when running at a speed of 0.15 mile per minute.

(II) The chaining together of multiple function processes (***function composition***) is useful for directly relating two quantities that are not obviously directly related through known information. Two function processes can be composed if the output quantity of the first function is the same as the input quantity of the second function. The chaining together of the two function processes is a way of creating a new function that defines the output quantity of the second function "in terms of" the input quantity of the first function. In this example, we say that the number of calories Mila burns when running 0.15 mile per hour, *c*, is a function of the number of minutes that Mila ran, *t*.

*2. a. Draw a diagram of a square and label the side lengths x (measured in inches). Visualize the square growing and shrinking as the value of x changes and think about how the square's perimeter length (in inches) compares to the side length.

b. Define a function g that inputs the square's perimeter length, P, and outputs the square's side length, x (both measured in inches).

c. How does the square's side length change as the perimeter changes from 6 inches to 20 inches? Calculate this value **and** represent it using function notation.

d. Define a function h that inputs the square's side length (in inches), x, and outputs the square's area (in square inches), A.

e. Using the functions in parts (b) and (d), determine the areas of squares that have each of the following perimeters: 24 inches, 60 inches, and P inches.

Reflecting on Exercise #2, you should notice that we defined two functions and that the output quantity of one function matched the input quantity of the second function. Thus, when given a square's perimeter (in inches), we follow a two-step process to find that square's area (in square inches).

- **Step 1:** Use g, which inputs the square's perimeter length and outputs its side length (both in inches).
- **Step 2:** Use h, which inputs that side length (in inches) and outputs the square's area (in in^2).

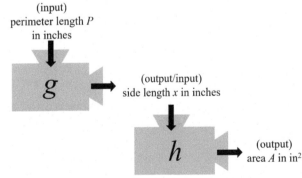

Now, we can also combine these steps into a <u>single</u> function that links the two processes together. **That is, we can define a function *f* that inputs the square's perimeter length in inches and outputs its area in square inches.**

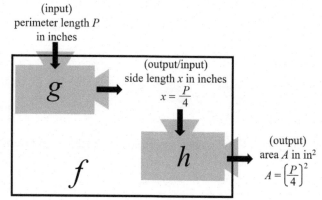

Let's write a formula defining *f*. We'll start with descriptions and move toward a symbolic representation.

$f(\text{perimeter length}) = \text{area}$

$f(\text{perimeter length}) = (\text{side length})^2$ • a square's area is product of its length and width

$f(\text{perimeter length}) = \left(\frac{\text{perimeter length}}{4}\right)^2$ • a square's side length is $\frac{1}{4}$ of its perimeter length

$$\boxed{f(P) = \left(\tfrac{P}{4}\right)^2}$$

So $f(P) = \left(\frac{P}{4}\right)^2$ where $A = f(P)$. Function *f* is a ***composite function*** – it is formed by uniting the processes of two functions (in this case, *g* and *h*) to form a single function.

Function Composition

Function composition is the process of chaining together two (or more) function processes by linking the output quantity of one function with the input quantity of another function.

The function created by doing this is called a ***composite function***.

---

3. Use the table of values to evaluate or solve each of the following.

    *a. $g(f(-1))$ *b. $f(g(3))$

        O   4

x	$f(x)$	$g(x)$
−2	0	5
−1	3	3
0	4	2
1	−1	1
2	6	−1
3	−2	0

    c. $f(f(3))$ d. $g(g(0))$

        O   ‑1

    e. If $f(g(x)) = 3$, then what must be the value of *x*?

        $x = 2$

*4. The given graphs show two functions, *f* and *g*. Function *g* takes as its input the forecasted high temperature in degrees Fahrenheit and outputs the expected attendance at a neighborhood carnival. Function *f* takes as its input the number of people attending the carnival and outputs the total expected revenue the carnival earns based on that attendance.

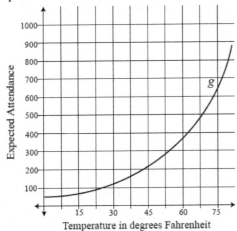
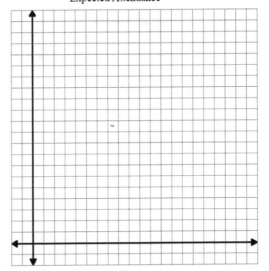

a. If the forecast predicts clear skies and a high temperature of 45°F, what is the expected revenue from the carnival today? What if the forecast predicts a high temperature of 75°F?

b. Function *h* inputs the forecasted high temperature (in °F) and outputs the expected revenue (in dollars). On the given axes, plot at least 5 coordinate points representing input/output pairs for *h*.

In Exercise #4, we used outputs of *g* as inputs to *f* in order to determine the expected revenue given a forecasted high temperature. The notation for representing ***expected revenue*** given the forecasted high temperature should seem very logical.

    expected revenue

f(expected attendance)   • *f* inputs expected attendance and outputs expected revenue

$f\big(g(\text{high temp °F})\big)$ • *g*(high temp °F) represents the expected attendance given the high temp °F

$f\big(g(x)\big)$ • Let *x* represent the high temperature in °F

So $f\big(g(x)\big)$ represents the expected revenue (in dollars) given *x*, the forecasted high temperature in degrees Fahrenheit. Evaluating an expression like $f\big(g(15)\big)$ is like following the order of operations. We begin with the inside function. *Note that we used the graph to estimate each value.*

$f\big(g(15)\big)$

    $f(75)$ • the expected attendance is 75 when the high temperature is 15°F

    200   • the expected revenue is $200 when the expected attendance is 75 people

Note that $f(g(x))$ represents the function's output value, but it is not the function's *name* (just like $f(x)$ is the output of f, not the function name). We use the notation $f \circ g$ to name the composite function.

Function Composition Notation

If f and g are functions, then $f \circ g$ is the name of the composite function formed by chaining together the two processes (where g is the "inside" function). The process of function composition involves:

1. Inputting a value into g and producing an output value.
2. Using that output value of g as an input to f to get another output value. The output value from f is the output value for the composite function.

If $f \circ g$ is the name of the composite function, then $f(g(x))$ represents the composite function's output value. *Note that we sometimes condense the notation by just giving this new composite function its own name, like h. So, we could define a function h to be the composite function $f \circ g$ by saying $h(x) = f(g(x))$.*

---

5. Use the following functions to answer the questions: $f(x) = \sqrt{x+3}$, $g(x) = 2x+9$, $h(x) = \frac{x}{4}$.

   *a. Evaluate $f(g(2))$.

   b. Evaluate $h(f(61))$.

   *c. Function m is defined as $m(x) = g(h(x))$. Write the formula for m.

   d. Function p is defined as $p(x) = g(f(x))$. Write the formula for p.

1. Use the graph provided to evaluate or solve each of the following. *Approximations are acceptable.*

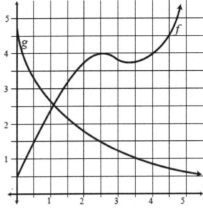

*a. $f(g(2.5))$ *b. $g(f(4))$ c. $g(g(3.5))$

    3.1   0.8   2.6

d. Determine the value(s) of x such that $f(g(x)) = 2.5$.

      $f(x) = 2.5$
        $x = 1.1$ or 1.2
      $g(x) = 1.1$
        $x = 3.4$

*2. A pebble is thrown into a lake and the splash creates a circular ripple. The circular ripple's radius length increases at a constant rate of 7 cm per second. Your goal is to represent the area (in square centimeters) inside the ripple in terms of the number of seconds since the pebble hit the water.

    a. Draw a picture of the situation and label the quantities. Discuss in a group what processes need to be carried out to determine the area inside the ripple when the number of seconds since the pebble hit the water is known.

    b.   i. As the time since the pebble hit the water increases, how does the circular ripple's radius length change?

        ii. As the circular ripple's radius length increases, how does the area inside of the ripple change?

       iii. As the time since the pebble hit the water increases, how does the area inside of the ripple change?

    c. Define a function f that represents the circular ripple's radius length (in centimeters), r, in terms of the number of seconds since the pebble hit the water, t.

d. Define a function g that represents the area inside the ripple (in square centimeters), A, in terms of the circular ripple's radius length (in centimeters), r.

e. Describe the process of evaluating $g(f(4.5))$.

f. In the expression $g(f(4.5))$, describe the quantity and unit of measurement associated with each of the following.

	Quantity	Unit of measurement
i) 4.5	_____	_____
ii) $f(4.5)$	_____	_____
iii) $g(f(4.5))$	_____	_____

g. Define a function h that determines the area inside the circular ripple (measured in square centimeters), A, as a function of the time since the pebble hit the water (measured in seconds), t.

h. Compute the value of $h(4)$ and the value of $g(f(4))$. What do you observe?

3. a.   If another pebble hits the water and its circular ripple's radius length increases at a constant rate of 3 cm per second (instead of 7 cm per second), write an *expression* that represents the area inside the ripple (in square cm) in terms of *t*, the number of seconds since this pebble hit the water.

   b.   A third pebble hits the water and its circular ripple's radius length increases at a constant rate of 5 cm per second. Define a function *j* that determines the area inside the ripple (in square cm), *A*, in terms of the number of seconds since this pebble hit the water, *t*.

*4.   a.   What is the area of a circle with a circumference 15 feet long? Describe the multiple step process you used to determine your answer.

   b.   Define a function *h* that determines a circle's area in terms of its circumference.

   c.   Use the function *h* you defined in part (b) to determine a circle's area when its circumference is 15 feet long. How does your answer compare to your answer in part (a)?

*5. *Recall the following context from Module 2 Investigation 4:* Function *g* takes as its input the forecasted high temperature in degrees Fahrenheit and outputs the expected attendance at a neighborhood carnival. Function *f* takes as its input the number of people attending the carnival and outputs the total expected revenue the carnival earns based on that attendance.

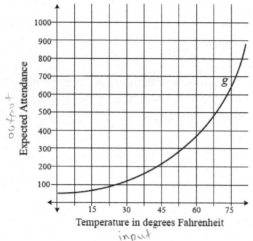
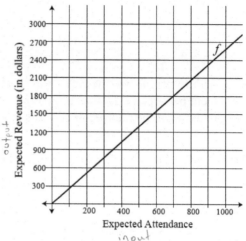

a. Does $f(g(70))$ represent a quantity's value in this context? If so, describe what its value represents. If not, explain why it doesn't represent a quantity's value in this context.

$g(70) \approx 500$ $f(500) \approx 1300$

1300 represents an expected $1300 of carnival revenue

$70 = 70° F$

$g(temperature) = f(attendance) = revenue$

b. Does $g(f(70))$ represent a quantity's value in this context? If so, describe what its value represents. If not, explain why it doesn't represent a quantity's value in this context.

$f(70°) = revenue$ and $g(revenue)$ isn't possible
not a quantity, doesn't work

c. Let's define a new function *k* that is the composition of *f* and *g*; that is, $k(x) = f(g(x))$. Explain what the equation $1800 = k(x)$ represents, then find the value of *x* that satisfies the equation.

$k(x) \rightarrow f° \rightarrow$ attendance $\rightarrow$ revenue
$\quad\quad\quad g(x) \quad\quad f(x)$

$k(x) \leftarrow \sim 77°F \leftarrow 700 \leftarrow \$1800$
$\quad\quad\quad\quad\quad\quad\quad$ people

$x = 77° F$ (roughly)

6. A farmer has 250 feet of fencing to create a rectangular pen for holding animals.
   a. Draw a picture of the situation and label the relevant quantities.

   b. If w represents the pen's width (in feet) and l represents the pen's length (in feet), then the pen's perimeter length is $2w + 2l$ feet, which must always total 250 feet (that is, $2w + 2l = 250$). Solve this equation for w.

   c. If the pen is 85 feet long, what is its width? What is its area? What if the pen is 46.5 feet long?

   d. For any value of w, what must be the corresponding value of l? The corresponding area?

   e. Define a function that expresses the pen's area (in square feet), A, as a function of w, the pen's width (measured in feet).

7. Let $w = 312q - 100$ and let $b = 21q^2$. Define a function h that represents w in terms of b. *Assume that b and q must be positive numbers.*

*1. The formula $P = 4s$ represents a square's perimeter length (in inches), P, in terms of the square's side length (in inches), s.

   a. Given a square with a perimeter length of 20 inches ($P = 20$), what is the square's side length? Demonstrate the reasoning you used to determine your solution.

   b. When we solve the equation $20 = 4s$ for s, this is one instance of reversing the formula's process. (*That is, instead of using the formula and a value of s to determine the corresponding value of P, we are using a formula and a value of P to determine the corresponding value of s.*) Use this same reversed process to determine the square's side length s for each of the following perimeter lengths.

     i. $P = 36$ ii. $P = 22$ iii. $P = 100$

   c. Generalize the process you engaged in when answering parts (a) and (b) by writing a formula that represents a square's side length, s, in terms of its perimeter length, P.

   d. How are the formula $P = 4s$ and the formula you generated in part (c) related? Do they express the same relationship between a square's side length and its perimeter length? Explain.

We have seen that relationships where every value of the independent quantity produces exactly one value of the dependent quantity are functions and can be represented using function notation. Recall that function notation allows us to reference a specified function relationship by name. We can also discuss a function output (dependent) value, such as $f(2)$, without having to determine its value.

Function Relationships

If $f(s) = 4s$, then a value of a square's side length (in inches) is input (or put in) to f. Applying the function rule, $4s$, determines a value for the square's perimeter length, $f(s)$, in inches.

We can use the image of a "function machine" to evaluate $f(8)$. We input a particular value of 8 in for s, and apply the rule of f (that is, we multiply 4 times 8) to obtain a value for the perimeter, $f(8) = 32$ inches.

---

2. Using the example in the "Function Relationships" box, explain how $f(12)$ is processed by f. (*Hint: Use the given image, which explains how f processes a value for the input quantity to produce the corresponding value of the output quantity.*)

A Function's Inverse

If $g(P) = \dfrac{P}{4}$, then a value of a square's perimeter length (in inches) is input (or put in) to g. Applying

the function rule, $\dfrac{P}{4}$, determines a value for the square's side length, $g(P)$, in inches.

If g is a function that undoes the process of the function f, we call g **the inverse function of f**. By convention, we represent the inverse function of some function f by writing f^{-1}, so we

can rename g as f^{-1} and thus $f^{-1}(P) = \dfrac{P}{4}$.

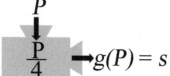

---

*3. a. Using the example in the definition boxes, evaluate $g(32)$ and explain how the input value 32 is processed (or operated on) by g.

b. What do you notice about how the input and output quantities for f and g are related?

c. Use function composition to evaluate the following:
    i) $g(f(5))$ ii) $f(g(25))$

    iii) $g(f(9))$ iv) $f(g(36))$

d. T or F: For the functions f and g defined in the definition boxes, f undoes the process of g, and g undoes the process of f.

---

e. How else could we have named function g in this context?

f. Discuss in groups or as a class how a function f and its inverse function f^{-1} are related in terms of:

i. the processes (or operations) they perform.

ii. their input quantities (and variables) and output quantities (and variables).

*4. When traveling outside the United States, it is often useful to be able to determine the temperature in Fahrenheit degrees given the temperature in Celsius degrees. The standard formula for determining the temperature in degrees Fahrenheit, F, when given the temperature in degrees Celsius, C, is $F = \frac{9}{5}C + 32$. $12°C = 53.6°F$ $0°C = 32°F$ $20°C = 68°F$

$72°F = 22.\overline{2}°C$
$100°F = 37.\overline{7}°C$
$58°F = 14.\overline{4}°C$
$42°F = 5.56°C$
$85°F = 29.\overline{4}°C$
$22.22°C = 71.996°F$
$53.6°F = 12°C$

a. Determine the formula that represents the temperature in degrees Celsius, C, in terms of the same temperature measured in degrees Fahrenheit, F. What is the independent quantity for this formula? What is the dependent quantity for this formula?

$C = \frac{5}{9}(F - 32)$
OR
$C = \frac{F - 32}{1.8}$

b. Define a function g that represents the temperature in degrees Fahrenheit, F, in terms of the same temperature measured in degrees Celsius, C.

$g(C) = \frac{9}{5}C + 32$

c. Define a function h that represents the temperature in degrees Celsius, C, in terms of the same temperature measured in degrees Fahrenheit, F.

$h(F) = \frac{5}{9}(F - 32)$ OR $h(F) = \frac{F - 32}{1.8}$

d. What does $g(C)$ represent? What does $h(F)$ represent?

$g(C) =$ forenheit

$h(F) =$ celsius } but outputs aren't in the equation

e. Determine the values of:

i. $g(100)$ ii. $h(212)$ iii. $h\big(g(100)\big)$ iv. $g\big(h(212)\big)$

$212°F$ $100°C$ $100°C$ $212°F$

When two functions represent the same relationship between covarying quantities, but their input and output quantities are reversed, we say that one function is the ***inverse of the other function***. In Exercise #4, the function *f* is the inverse of function *g*, and the function *g* is the inverse of *f*.

A Function's Inverse (*Continued*)

A function and its inverse relation represent the same relationship between two covarying quantities but with the independent (input) and dependent (output) quantities reversed.

If function *g* is the inverse of function *f*, then *g* undoes the process of *f*, and *f* undoes the process of *g*.

The idea of function inverse shouldn't be entirely new. When you rewrite a two-variable formula to solve for the alternative (or other) variable, you are defining the inverse relationship.

5. Solve each of the following formulas for the other variable.

   a. $y = 2x - 8$ for x

   b. $w = \dfrac{r + 19}{4}$ for r

   c. $n = \dfrac{2x + 6}{x}$ for x

Mathematicians developed a special function name to communicate that two functions are inverses.

Inverse Function Notation   "*f inverse*"

Given a function *f*, its inverse (if also a function) is named f^{-1}. Since the input and output quantities are reversed in these two functions, we have that, if $y = f(x)$, then $x = f^{-1}(y)$.

The use of "–1" here is not an exponent. It's just part of the function name.

REMINDER: According to the rules of exponents, for any non-zero real number *n*, $n^{-1} = \frac{1}{n}$. Even though we are using similar notation here, it is ***not true*** that for any function *f* we have $f^{-1} = \frac{1}{f}$. You have to use context clues to understand the notation and recognize when f^{-1} is referring to the name of a function.

*6. Define the inverse function for each of the following functions.

   a. $f(x) = 3x - 7$ with $y = f(x)$

   $$f^{-1}(y) = \frac{y + 7}{3}$$

   b. $j(r) = \dfrac{2r - 5}{6}$ with $w = j(r)$

   $$w6 = 2r - 5$$
   $$+5 \qquad +5$$
   $$\frac{w6 + 5}{2} = \frac{2r}{2}$$
   $$x = \frac{w6 + 5}{2}$$
   $$j^{-1}(w)$$

*7. A ball is dropped off the roof of a building. The given graph of function f represents the ball's height above the ground (in feet), h, in terms of the number of seconds since the ball was dropped, t.

a. What does $f(4)$ represent in this situation?

Input *Output*

The ball's height 4 seconds after it was dropped in terms of seconds

height in terms of time

b. What does $f^{-1}(4)$ represent in this situation?

time in terms of height

Time since ball was dropped when ball is 4 ft above ground

The ball is 4 ft above the ground some amount of time after it's been dropped

c. Estimate the value of $f^{-1}(250)$ and explain what the value represents in this context.

≈2.75

The ball is 250 feet above the ground ≈2.75 seconds after being dropped.

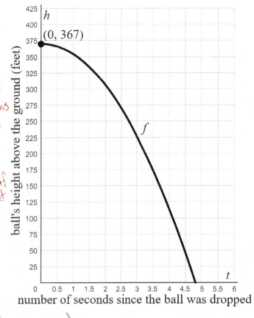

ball's height above the ground (feet)

(0, 367)

h

f

t

number of seconds since the ball was dropped

d. Estimate the solution to the equation $f(t) = 75$ for t, then demonstrate how to represent this value using inverse function notation.

8. Use the given table to evaluate the following.

*a. $f(0)$ *b. $g^{-1}(3)$ c. $g^{-1}(0)$
    4   -1   3

d. $f^{-1}(-1)$ *e. $f^{-1}(g(2))$ *f. $g^{-1}(f^{-1}(6))$
    1   1   0

x	$f(x)$	$g(x)$
−2	0	5
−1	3	3
0	4	2
1	−1	1
2	6	−1
3	−2	0

Composition of Inverses

Inverse functions represent the same relationship between two quantities, but with the input and output quantities reversed. We might say that inverse functions "undo" each other. If $f(a) = b$ then it must be that $f^{-1}(b) = a$ (assuming that f's inverse is a function and assuming that a is an element in f's domain and b is an element of its range).

Thus,

$$f^{-1}(f(a)) = f^{-1}(b) \qquad \bullet \text{ since } f(a) = b$$
$$= a \qquad \bullet \text{ since } f^{-1}(b) = a$$

$$f(f^{-1}(b)) = f(a) \qquad \bullet \text{ since } f^{-1}(b) = a$$
$$= b \qquad \bullet \text{ since } f(a) = b$$

The composition of two inverses will always produce the original input value for the "inside" function (in either order). In fact, this is one way to test if two functions are inverses.

9. Your classmate believes that each of the following pairs of functions are inverses. Is your classmate correct? Verify by creating the formula for $g(f(x))$ and simplifying to see if $g(f(x)) = x$.

   *a. $f(x) = 2(x+1) - 8$ and $g(x) = \dfrac{x+8}{2} - 1$ b. $f(x) = \dfrac{3x+7}{4}$ and $g(x) = \dfrac{4}{3}x - 7$

*10. We are given that a function f exists, that its domain and range include all real numbers, that its inverse, f^{-1}, is also a function, and that $f(7) = 2$. Three of your classmates provided explanations about the relationship between f and f^{-1}. Which of them (if any) are correct? Explain your thinking.

   Calvin: "f^{-1} inverts f, so $f^{-1}(7) = \dfrac{1}{2}$."

   Brecklyn: "f^{-1} inverts f, so $f^{-1}(2) = \dfrac{1}{7}$."

   Myshelle: "f^{-1} inverts f, so $f^{-1}(f(7)) = 7$."

*11. Given the function f defined by $f(x) = x^2$ with $y = f(x)$, complete the following.
   a. Evaluate $f(2)$ and $f(-2)$.

   b. Explain why reversing the process for f produces a relationship that is not a function. (*Hint: What is the definition of a function?*)

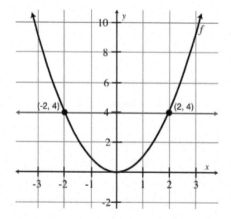

   c. Consider $g(x) = x^2$ (with $y = g(x)$), where the domain of g is restricted so that $x \geq 0$.
   i. Explain why $g's$ inverse is a function.

   ii. Write the formula for g^{-1} and give its domain and range.

In Module 1, Investigation 5 we introduced the idea of *average rate of change* in the context of a function formula and graph.

Recall that we determined the average rate of change of a function on a specified interval of a function's domain by comparing the change in the dependent variable's value over the interval to the change in the independent variable's value over that interval. Our goal was to determine what *constant rate of change* would achieve the same net change in the function's dependent quantity over the same interval of the independent quantity.

We then used this average rate of change (which is also a constant rate of change) value to make estimates, such as determining the amount of time needed to travel some distance at this rate of change.

Average Rate of Change (over some interval)

The *average rate of change* of a function over the interval of the domain from $x_1 = a$ to $x_2 = b$ is the constant rate of change over that interval that produces the same net change in the function's output values, from $y_1 = f(x_1)$ to $y_2 = f(x_2)$.

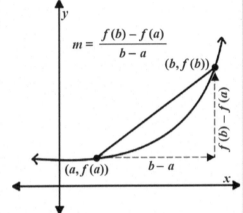

Graphically, the *average rate of change* of a function on an interval of a function's domain is represented as the slope of the line passing through the points at the beginning and the end of the interval. Thus, we use the slope formula to calculate the average rate of change on that interval:

$$m = \frac{y_2 - y_1}{x_2 - x_1} = \frac{f(b) - f(a)}{b - a}$$

The average rate of change is a common tool for estimating the rate of change of a function over some interval of the function's domain in which the rate of change is not constant.

In this investigation, we define the average rate of change using function notation and determine a function's average rate of change on domain intervals for the purposes of explaining how a function's rate of change varies over successive intervals of the function's domain and approximating an object's rate of change over some interval.

1. Kylie left her dormitory this morning at 8:20 am and walked 384 meters to her first class over the next 6 minutes.

   - During the first minute, she traveled 70 meters.

   - During the second minute, she traveled 90 meters.

   - After 2 minutes of walking, she gradually slowed down as she walked for another 65 meters before stopping at an intersection 3 minutes since leaving her dorm.

   - After waiting at the intersection for 1 minute, she spotted a friend and immediately started walking.

   - Over the course of the next minute, she gradually increased her speed and caught up with her friend after walking 95 meters from the intersection.

   - She continued walking with her friend at a constant speed for 1 minute, reaching her class 6 minutes after leaving her dorm.

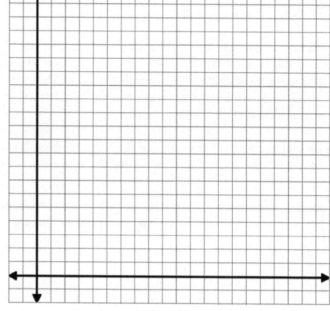

   a. Construct a "rough sketch" of a graph of function f, representing Kylie's distance walked from her dorm (in meters) in terms of the amount of time (in minutes) since she left her dorm. Label and scale your axes and plot each of the following points: $(0, f(0))$, $(1, f(1))$, $(2, f(2))$, $(3, f(3))$, $(4, f(4))$, $(5, f(5))$, and $(6, f(6))$.

   b. Determine the value of $f(6) - f(2)$ and represent this value on your graph. Explain what this value represents in the context of this problem.

   c. Connect the points $(2, f(2))$ and $(6, f(6))$ with a straight line, then use function notation to represent the line's slope.

   d. Determine Kylie's average speed over the interval from 2 to 6 minutes since she left her dorm and then explain what your solution represents in this context.

e. How is your answer to part (d) represented on the graph?

*2. A car is driving away from a crosswalk. The distance d (in feet) of the car from the crosswalk t seconds since the car started moving is given by the function $f(t) = \frac{1}{2}t^2 + 3$ with $d = f(t)$.

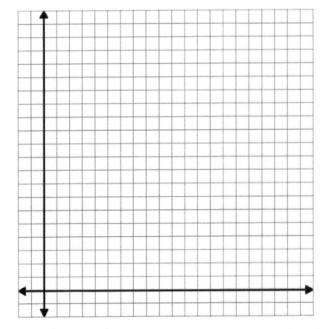

a. As the number of seconds since the car started moving increases from 2 seconds to 6 seconds, what is the change in the car's distance from the crosswalk?

b. Graph f on the given axes, then illustrate how to represent each of the following.

   i. The increase in t from 2 to 6 seconds since the car started moving.

   ii. The corresponding change in the car's distance from the crosswalk as the value of t increased from 2 to 6 seconds.

c. True or False: The car travels at a constant speed as the value of t increases from 2 to 6 seconds. Discuss with your group or as a class and explain your conclusion. (*Hint: It may help to think about how far the car travels in the 1ˢᵗ second of the interval compared to how far it travels in the 2ⁿᵈ second of the interval.*)

d. Since the car is not traveling at a constant speed, it can be challenging to estimate its speed over any given interval. It is a common practice to <u>estimate</u> the car's speed during an interval by *pretending* its speed was constant and asking, "What constant speed would have resulted in the car traveling the same distance (as the distance that the car actually traveled) over this time interval?"

   i. Draw a line passing through the points $(2, f(2))$ and $(6, f(6))$. Determine the value of $\dfrac{f(6) - f(2)}{6 - 2}$ and explain what this value represents in the context.

   ii. Plot the point $(9, f(9))$ and then draw a line through the points $(6, f(6))$ and $(9, f(9))$.

   iii. Determine the average speed of the car on the interval $6 \leq t \leq 9$ and explain what the value represents in this problem context.

   iv. Is the car speeding up, slowing down, or traveling at a constant speed on the interval from $t = 2$ to $t = 9$ seconds? Explain.

   v. <u>If</u> the car traveled at the constant speed you found in part (iii), how far would it travel during the next 0.5 second, from $t = 9$ to $t = 9.5$ seconds?

vi. How might you use the idea of average rate of change to get a better approximation of the car's speed at about 9 seconds since the car began moving?

e. True or False: The line passing through the points $(6, f(6))$ and $(9, f(9))$ represents the graph of the car's actual distance (in feet) from the crosswalk as t increases from 6 to 9 seconds. Explain.

As we mentioned, it's common to use the behavior of a linear function over an interval as a way to approximate the behavior of functions that don't have constant rates of change. The constant rate of change we get by doing this is called the ***average rate of change*** of the function over that interval.

Average Rate of Change

The average rate of change of a function on an interval from $x = a$ to $x = b$ is the constant rate of change that would produce the same net change in output values over that interval.

---

*3. Given $f(x) = \sqrt{x+3}$ and $g(x) = \dfrac{10}{x}$, do the following.

a. Find the average rate of change of f over the interval from $x = 1$ to $x = 6$ and then again over the interval from $x = 3$ to $x = 10$. Explain what your solutions represent.

b. Find the average rate of change of g over the interval from $x = 1$ to $x = 4$ and then again over the interval from $x = 5$ to $x = 10$. Explain what your solutions represent.

---

*4. Use function notation to represent the average rate of change of a function f over the interval from $x = 4$ to $x = 7$.

*5. Use function notation to represent the average rate of change of a function g over the interval from $x = -3$ to $x = 0$.

*6. Use function notation to represent the average rate of change of a function h over the interval from $x = n$ to $x = n + 2$.

The Difference Quotient

The average rate of change of a function f on any interval h units long is represented by *the difference quotient*:

$$\frac{f(x+h) - f(x)}{(x+h) - x} \quad \text{or} \quad \frac{f(x+h) - f(x)}{h}$$

I. THE BOX PROBLEM AND MODELING RELATIONSHIPS (TEXT: S1)

1. Consider the volumes of various open-topped boxes that can be constructed by cutting four equal-sized square corners from a 14-inch by 17-inch sheet of cardboard and folding up the sides.
   a. Create an illustration to represent the situation and label the relevant quantities.
   b. What are the box's dimensions (length, width, and height) if the length of the side of the square cutout is 0.5 inch? 1 inch? 2 inches?
   c. Let x represent the length of the side of the square cutout (in inches); let w represent the width of the box's base (in inches); let l represent the length of the box's base (in inches); and let V represent the box's volume (in cubic inches). Complete the given table of values.

x	w	l	V
0			
4			
9			
		2	
	0		

   d. Define a formula to represent the height of the box in terms of the length of the side of the square cutout. Use the variables defined in part (d).
   e. Define a formula to represent the box's volume in terms of the length of the side of the square cutout. Use the variables defined in part (d).
   f. Use the formula created in part (g) to determine the change in the volume of the box as the length of the side of the square cutout increases from 0.5 inches to 1 inch and from 1 inch to 1.5 inches. Why are these values not the same?
   g. Use your graphing calculator to approximate the maximum volume of the box rounded to the nearest tenth of a cubic inch. What cutout length produces this maximum volume?

2. Consider the exterior surface areas for various open-topped boxes that can be constructed by cutting four equal-sized square corners from a 9-inch by 13-inch sheet of cardboard and folding up the sides. (*Note: The exterior surface area is the total area of the box's sides and bottom that face "away" from the box's interior.*)
   a. Create an illustration to represent the situation and label the relevant quantities.
   b. Complete the given table of values.
   c. Define a formula that represents the total exterior surface area based on the length of the side of the square cutout. Define any variables you use.

Cutout length (inches)	Box height (inches)	Box Length (inches)	Box Width (inches)
0.5			
1			
2			

   d. There are two ways to think of determining a box's exterior surface area: (1) by adding up the areas of each piece of the box; and (2) by subtracting the areas of the four cutouts from the area of the initial sheet of cardboard.
      i. Which method did you use in defining your formula in part (c).
      ii. Define a formula to compute the box's exterior surface area using the other method. Compare the two formulas. Are they equivalent (that is, do they always produce the same result)? Justify your answer algebraically.

3. Find the value of $\dfrac{(9+x)-x^2}{2x+7}$ when $x = 3$.

4. Find the value of $\dfrac{3x^2 + x - 2(3x+5)}{2x+9}$ when $x = 8$.

5. Find the value of $\dfrac{(x+4)^3 - 4y + 6xy}{-5x + 6y}$ when $x = 2$ and $y = 0.5$.

6. Find the value of $y\sqrt{x+4} - 3x - 7y + \dfrac{4x}{8}$ when $x = 12$ and $y = -4$.

7. Find the value of $\left(2(x-7)\right)^2 - \dfrac{y}{6}$ when $x = -2$ and $y = 54$.

8. An open-topped box is constructed by cutting four equal-sized square corners from a 10-inch by 13-inch sheet of cardboard and folding up the sides.

    a. Define a formula to represent the box's volume in terms of the length of the side of the square cutout. Define any variables you use.

    b. Determine the volume of the box when the length of the side of the square cutout is:
        i. 1 inch   ii. 2.5 inches   iii. 5 inches

    c. Use the graph to approximate the value(s) of the length of the side of the square cutout when the volume is:
        i. 0 in^3 ii. 100 in^3 iii. 108.4 in^3

    d. As the length of the side of the square cutout increases from 0 inches to 1 inch, by how much does the volume change? Illustrate this change on the graph.

    e. For what interval(s) of the cutout length is the box's volume increasing? Decreasing?

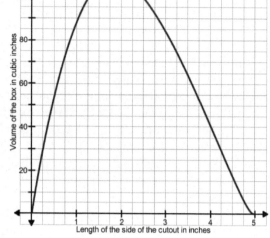

9. *Windows, Inc. manufactures specialty windows. One of their styles is in the shape of a semicircle, as shown.* When a customer orders this window, they specify the length of the base of the window.

    a. Determine the frame's total perimeter length for a window with a base 4 feet long.

    b. Define a formula that represents the frame's total perimeter length in terms of the base's length. Define any variables you use.

    c. If the cost of the framing material is $12 per linear foot, what is the cost to frame a window with a base length of 4 feet?

    d. Suppose that your budget limits you to spending $250 on your window frame. What is the longest base length that you can afford to frame?

    e. The glass pane that will go inside of the window frame costs $23 per square foot. What is the area of a glass pane when the window base is 6 feet long?

    f. Define a formula that represents the area of the glass pane in terms of the window's base length. Define any variables you use.

    g. What is the cost of the glass pane when the base of the window is 6 feet long?

    h. Suppose that your budget limits you to spending $700 on your glass pane. What is the longest base length of the window that you can afford?

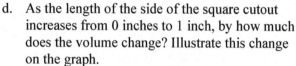

II. FUNCTION RELATIONS AND DOMAINS OF FUNCTIONS (TEXT: S1)

For Exercises #10-13, do the following.
    a. Determine the input quantity.
    b. Determine the output quantity.
    c. Determine if the relationship is a function. If it is, describe a possible domain and range.

10. a student's ID number in terms of their birth year

11. a student's birth year in terms of their student ID number

12. In a given apartment complex, the apartment number in terms of the number of people living at that apartment number.

13. In a given apartment complex, the number of people living at an apartment number in terms of the apartment number.

14. According to the information in the table, is the height of a person in terms of his/her driver's license number a function relationship? Explain your reasoning.

Person's Driver's License Number	DL4896578	DL9630654	DL8774309	DL1194096	DL2856498
Person's Height (inches)	71	63	64.5	71	68

15. According to the information in the table in Exercise #14, is a person's driver's license number in terms of his/her height a function relationship? Explain your reasoning.

16. For each of the given relations, determine whether y is a function of x based on the given information. Use the definition of a function to justify your answer.

a.

x	y
1	3
2	2
1.7	4.5
2.1	9
2	1.1
5	7

b.

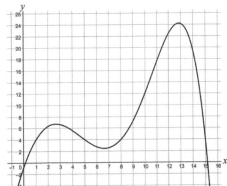

c. $y = x(8.5 - 2x)(11 - 2x)$

d. $y = 2^x$

e. A manufacturing plant produces bags of plaster weighing between 0 and 90 pounds. Bags weighing up to 50 pounds are marked with a "1" (indicating that the bags can safely be carried by one person), and bags weighing more than 50 pounds are marked with a "2" (indicating that the bags are heavy and should be carried by two people). Is the number marked on the bag, y, a function of the weight of the bag, x?

17. For each of the relations in Exercise #16 parts (a) through (d), determine whether x is a function of y. Justify your answers.

18. Create two tables that represent function relations between two quantities. Explain why one variable is a function of the other variable.

19. Create two tables that represent relations between two quantities that are NOT functions. Explain why one variable is not a function of the other variable.

Use the following functions for Exercises #20-24.

$$f(x) = 3x^2 + 5x - 7 \qquad h(x) = \frac{x}{5x - 10} \qquad k(x) = \sqrt{x + 2} \qquad n(x) = \frac{\sqrt{x - 7}}{x^2 - 4}$$

20. Evaluate each of the following expressions.
   a. $f(6)$ b. $h(-2)$ c. $k(9)$ d. $n(11)$

21. Explain the meaning of each of the following statements.
   a. $f(-3) = 5$ b. $k(17) \approx 4.36$

22. a. Explain what it means to find x such that $k(x) = 5$. (*Don't determine its value yet.*)
   b. Find the value(s) of x such that $k(x) = 5$.

23. Determine the domains of functions f and h.

24. Determine the domains of functions k and n.

Use the graphs of *f* and *g* for Exercises #25-29.

25. Evaluate each of the following expressions. *Approximations are okay.*
    a. $f(2)$ b. $f(0)$ c. $g(0)$ d. $g(4.5)$

26. Explain the meaning of each of the following statements.
    a. $f(1) \approx 0.24$ b. $g(2.5) \approx -3$

27. a. Explain what it means to find *x* such that $g(x) = 4$. (*Don't determine its value yet.*)
    b. Find the approximate value(s) of *x* such that $g(x) = 4$.

28. Determine the range of *f* (based on the information you know from the graph).

29. Determine the range of *g* (based on the information you know from the graph).

30. The expression $224 - 4x^2$ represents the exterior surface area of an open-topped box (measured in square inches) when the box is made with a square cutout of side length *x* (measured in inches) from a 14" by 16" sheet of paper.
    a. Define a function *g* that has the length of the square cutout, *x*, as input (measured in inches) and the exterior surface area of the corresponding box, *A*, as output (measured in square inches).
    b. Using function notation, represent the box's exterior surface area when the length of the side of the square cutout is 0.2 inch, 2.7 inches, and 4.1 inches. (That is, *do not* calculate the exterior surface areas. Instead, *represent* the surface areas determined by each length of the side of the square cutout.)
    c. What does the expression $g(1.3)$ represent?

31. The expression $\frac{4}{3}\pi r^3$ represents the volume of a sphere (measured in cubic inches) when the radius of the sphere is *r* inches.
    a. Define a function *h* that has the radius of the sphere, *r*, as input (measured in inches) and the volume of the sphere, *V*, as output (measured in cubic inches).
    b. Using function notation, represent the volume of the sphere when the radius of the sphere is 2.4 inches, 3.1 inches, and 5.2 inches. (That is, *do not* calculate the volumes. Instead, *represent* the volumes determined by each radius length.)
    c. What does the expression $h(19.6)$ represent?

III. USING AND INTERPRETING FUNCTION NOTATION (TEXT: S3, S4)

32. $g(x) = 224 - 4x^2$ represents the exterior surface area of an open-topped box (in square inches) when the box is made with a square cutout of side length, *x*, (measured in inches) from a 14" by 16" sheet of paper.
    a. What does the expression $g(1.5) - g(0.5)$ <u>represent</u> in this context? What is its value?
    b. What does the expression $g(6.25) - g(2.1)$ <u>represent</u> in this context? What is its value?

33. Function *h* represents a dairy farmer's cost (in dollars) to produce *b* gallons of milk. Explain what each of the following represents in this context.
    a. $h(18)$ b. $h(b) = 25$ c. $h(15) - h(9)$ d. $h(4.3) + h(6.4)$

34. Functions are commonly used by computer programmers to access databases. All databases that involve time have "reference times" (times they consider to be "0"). A certain database uses a function named g to access the population of Loveland, CO and uses January 1, 1990 as its reference time. So, $P = g(u)$ represents Loveland's population u years from January 1, 1990.
    a. Represent Loveland's population on 1/1/1982.
    b. Represent the change in Loveland's population from 1/1/1994 to 1/1/1997.
    c. Represent a population value 6 times as large as Loveland's population on 1/1/1972.
    d. What does $g(8) - g(-13)$ represent?
    e. What does $g(h + 3)$ represent, where h is some number of years?
    f. What does $g(h + 3) - g(3)$ represent, where h is some number of years?

35. Use function notation to represent each of the following.
    a. When the input to the function q is 10, the output is 12.
    b. The output of function h is c when the input value is 34.
    c. Function f represents the area of a square as a function of its side length s.
        i.   Define a formula for function f.
        ii. The area of a square is $25m^2$ when the side length is 5m.
        iii. Write an expression using the function f that represents how much larger the area of a square with side lengths of 6 m is than the area of a square with side lengths of 2 m.
    d. The circumference of a circle is π times as large as the circle's diameter d.
        i.   Define a function g to express the circumference of a circle in terms of its diameter.
        ii. Represent the circumference of a circle with a diameter length 3 times as large as a inches.

36. The given table provides values of a circle's circumference length (in cm), C, as a function of its radius length (in cm), r.
    a. What does the expression $f(1.5)$ represent? What is the numerical value of $f(1.5)$?
    b. What does the expression $f(2) - f(1.5)$ represent? What is the numerical value of $f(2) - f(1.5)$?
    c. When the circle's radius length changes from 1.5 to 2.5 cm, by how much does its circumference length change?
    d. Represent the solution to part (c) using function notation.

radius length (cm)	circumference (cm)
r	$C = f(r)$
1	6.283
1.5	9.425
2	12.566
2.5	15.708
3	18.850

37. Given the existence of some functions b and k, complete the following.
    a. Use function notation to represent the change in the output value of function k when the input value changes from 6 to 9.
    b. Use function notation to represent the sum of the output of the function k at an input of -9 and the output of the function b at an input of 6.12.

38. Use the following information to answer the questions that follow.
    - $C(t)$ represents the number of cats owned by people living in the U.S. t years since 2000.
    - $D(t)$ represents the number of dogs owned by people living in the U.S. t years since 2000.

    a. Represent, using function notation, the total number of cats and dogs, $P(t)$, owned by people in the U.S. t years since 2000.
    b. Represent, using function notation, how many _more_ cats than dogs were owned by people living in the U.S. t years since 2000.

39. Evaluate each of the following:

    a. $f(13)$ when $f(x) = 4.5 - 6x$
    b. $f(5)$ when $f(x) = \dfrac{2x^2 + (6 - 4x)}{9 - 3x}$

40. Evaluate each of the following:

    a. $f(x+2)$ when $f(x) = 4x^2 - 2x + 10$

    b. $h(2x)$ when $h(y) = \dfrac{y^3 - 2y^2 + 4}{2y}$

IV. FUNCTION COMPOSITION: CHAINING TOGETHER TWO FUNCTION PROCESSES (TEXT: S5)

41. Alejandra bought a new house and is planning to install landscaping. In her initial budget, she set aside $400 to purchase ¾-inch gravel to cover part of her yard. This size gravel covers approximately 130 square feet per ton. She has four different options for this size gravel depending on the quality of the gravel. In addition to the gravel cost, the company charges $89 for delivery.

¾-inch gravel type	Cost per ton (in dollars)
Grade A	$61.50
Grade B	$56.50
Grade C	$54.50
Grade D	$52.50

    a. How many tons of each grade of gravel can Alejandra afford if she must pay the $89 delivery charge?

    b. How many square feet of her yard can she cover with each grade of gravel? Explain your thinking.

42. Reggie, a college student, works part-time during the school year. He must budget his expenditures so that he has enough money at the end of the month to cover his rent, car payment, and insurance. Currently, Reggie can only afford $25 per week for gas.

    a. As the price of gas fluctuates, the amount of gas Reggie can purchase each week varies. Complete the table of values showing the number of gallons of gas Reggie can purchase with $25 at the given fuel prices.

    b. Reggie's car gets an average of 28 miles per gallon. How many miles can he drive in a week if he purchases 3 gallons of fuel? 4.18 gallons? g gallons?

    c. Explain how you can determine the number of miles Reggie can drive in a week if gas costs $3.899 per gallon and he has $25 to spend on gas.

    d. Complete the table of values showing the price of fuel (in dollars per gallon) and the number of miles Reggie can drive on $25 worth of gas.

    e. How many miles can Reggie drive on $25 worth of gas if gas costs p dollars per gallon?

Price of fuel (in dollars per gallon)	Number of gallons of fuel Reggie can purchase for $25
3.199	
3.499	
3.599	
3.799	
p	

Price of fuel (in dollars per gallon)	Number of miles Reggie can drive on $25 worth of gas
3.299	
3.449	
3.579	
3.839	

43. A ball is thrown into a lake creating a circular ripple with a radius length that increases by 7 cm per second. We want to express the area enclosed within the ripple in terms of the time elapsed since the ball hit the lake.

    a. Draw a diagram of the situation.

    b. Identify the quantities in the situation that vary in value and state what units you'll use to measure each of these quantities. Repeat for quantities with fixed values.

    c. As the amount of time (in seconds) since the ball hit the lake, t, increases over each of the given time periods, how does the ripple's radius length (in cm), r, change?
       i. from $t = 0$ to 3 seconds   ii. from $t = 4$ to 6 seconds   iii. from $t = 6$ to 6.5 seconds

    d. Define a function g that represents the value of r in terms of t.

    e. Define a function f that represents the area contained within the ripple (in square cm), A, in terms of t.

    f. Describe the meaning of $f(2.3)$ without performing any calculations. Then calculate the value of this expression and interpret its meaning.

44. When Jessie travels for work, he likes to go for a run in the morning to keep fit. Jessie has noticed that the elevation of the city he is visiting impacts how long he is able to spend running since there is less oxygen available at higher elevations. The given graphs model information about Jessie's exercise routines recorded over many business trips.

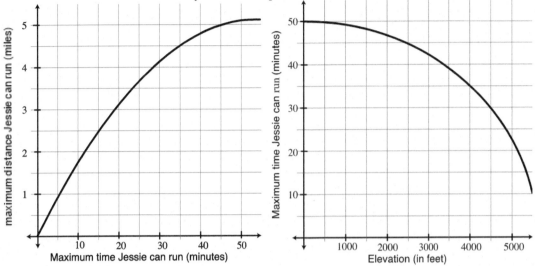

a. Describe how to determine the maximum distance Jessie can run if he is visiting a city with an elevation of 4000 feet.

b. Explain how you can determine the elevation of the city Jessie is visiting if he expects his maximum running distance to be 5 miles.

45. When hiring a contractor to add insulation to your attic, he provides you with information that you convert into tables specifying how input and output values are related for two functions, f and g.

Function f		Function g	
Number of bags of insulation applied, n	Depth of insulation (inches), $f(n)$	Depth of insulation (inches), d	Estimated annual heating/cooling costs (dollars), $g(d)$
11	2	15	$940
20	3.6	13	$1,000
28	5	11	$1,100
40	7.2	9	$1,250
50	9	7	$1,500
61	11	5	$1,900
70	12.6	4	$2,750
78	14	3	$3,600

a. Describe the meaning of $g(11)$.

b. Does the expression $f(g(11))$ have a real-world meaning in this context? If so, find the value and explain its meaning. If not, explain your reasoning.

c. Does the expression $g(f(50))$ have a real-world meaning in this context? If so, find the value and explain its meaning. If not, explain your reasoning.

d. Solve the equation $g(n) = 1900$ for n and explain the meaning of your solution.

e. Function T is defined as follows: $T(n) = g(f(n))$. Use the given diagram to describe what each of the parts (A, B, C, D, and E) of the definition represents.

$$T(n) = g(f(n))$$

46. You have a coupon for $10 off a purchase of $100 or more at Better Buys.
    a. Define a function C that determines the final cost of a purchase (in dollars) after applying the coupon if x is the original price (in dollars). Explain what the expression $C(215.83)$ represents.
    b. What are the domain and range of function C?
    c. This weekend, Better Buys has a customer appreciation sale that offers 5% off all purchases. Define a function T that determines the final cost of a customer's purchase after the sale discount if the original purchase price is p dollars. (Do not include the coupon in this function rule.)
    d. Better Buys will allow you to use both the sale and the coupon together. If you buy a television set that is priced at $1979.99, describe what the expression $C(T(1979.99))$ represents and determine its value.

47. An oil tanker crashed into a reef off the coast of Alaska, grounding the tanker and punching a hole in its hull. The tanker's oil radiated out from the tanker in a circular pattern. You are an engineer for the oil company and your task is to monitor the spill.
    a. Draw a diagram of the situation.
    b. What are the quantities that vary in value in this situation? Define variables to represent the values of these quantities. *Be sure to include the units of measure when defining the variables.*
    c. If the oil tanker was 105 feet from the spill's outer edge exactly 24 hours after the spill, how much area (in square feet) did the spill cover at that time?
    d. News reporters want to know how much oil has spilled at any given time. The only information available to you is satellite photos that include the spill's radius length (in feet). You also know that 7.5 gallons of oil creates 4 square feet of spill.
        i. Define a function g that represents the number of gallons of oil spilled in terms of the area of the oil spill. *Define any variables you use or use the variables you defined in part (b).*
        ii. Define a function f that represents the number of gallons of oil spilled in terms of the spill's radius length. *Define any variables you use or use the variables you defined in part (b).*
    e. Use function notation to represent the increase in the number of gallons spilled as the spill's radius length increases from 121 feet to 152 feet.

48. Given $f(x) = 2x+1$ and $g(x) = x^2 + 2x + 1$, complete the following.
    a. Fill in the first two tables and use these two tables to complete the third table.
    b. Describe how you used the first two tables to determine the values of the third table.

x	$f(x)$
0	
1	
1.5	
2	
5	

x	$g(x)$

x	$g(f(x))$
0	
1	
1.5	
2	
5	

49. Use the given tables to answer the following questions.
    a. Evaluate the following expressions:
        i. $g(f(-2))$ ii. $f(g(0))$ iii. $g(g(4))$
        iv. $f(g(2))$ v. $g(f(-1))$ vi. $f(f(0))$
    b. Solve the equation $g(f(x)) = -10$ for x.
    c. Solve the equation $f(g(x)) = 16$ for x.

x	$f(x)$
−4	16
−3	13
−2	6
−1	0
0	−4
1	−7
2	−9
3	−10

x	$g(x)$
−8	−20
−6	−17
−4	−10
−2	−4
0	−1
2	2
4	6
6	11

50. The functions f, g, and h (given below) are used to define the functions s, r, and v. Re-write the definitions of s, r, and v so that their definitions do not involve function composition.

$$f(x) = \frac{2x+1}{x-3} \qquad g(x) = x^2 - 7 \qquad h(x) = -3x + 2$$

    a. $s(x) = f\big(g(x)\big)$ b. $r(x) = h\big(h(x)\big)$ c. $v(x) = g\big(h(x)\big)$

51. Use the words *input* and *output* to describe how you would evaluate each of the following composite functions. Assume that g and h are both functions.

    a. $h\big(g(2)\big)$ b. $g\big(g(3)\big)$ c. $g\big(h(x+1)\big)$

52. The functions g, h, and k are defined below. Use these functions to evaluate each function composition or solve the given equation.

$$g(x) = x + 5$$

x	h(x)
−1	−13.4
0	− 7.3
2	− 4.4
4	3
5	6.8
6	15
8	22

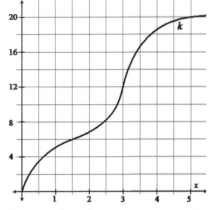

    a. $k\big(g(0)\big)$ b. $h\big(k(1.5)\big)$ c. $g\big(h(2)\big)$ d. $k\big(h(4)\big)$

    e. Solve the equation $h\big(g(x)\big) = 22$ for x.   f. Solve the equation $g\big(k(x)\big) = 17$ for x.

53. Use the graphs below to answer the questions that follow.

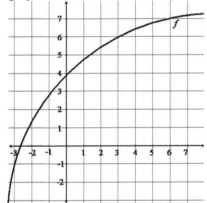
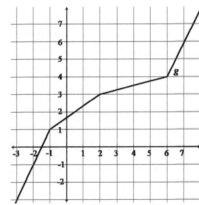

    a.   Approximate the value of each of the following compositions.

        i. $f\big(f(3)\big)$ ii. $g\big(f(6)\big)$ iii. $g\big(g(-2)\big)$ iv. $g\big(f(-3)\big)$

    b.   Find the value of x that satisfies each of the following equations.

        i. $f\big(g(x)\big) = 6$ ii. $g\big(f(x)\big) = 4$

54. For each of the given functions, redefine the function in terms of two new functions, f and g, using function composition and function arithmetic. *For example,* $f(x) = (2x)^3$ *can be defined as*

    $f(x) = h\big(g(x)\big)$ *if* $g(x) = 2x$ *and* $h(x) = x^3$.

    a. $h(x) = 3(x - 1) + 5$ b. $m(x) = (x + 4)^2$ c. $k(\underline{x}) = (x + 2)^2 + 3(x + 2) + 1$

    d. $j(x) = \sqrt{x - 1}$ e. $p(x) = \dfrac{500}{100 - x^2}$

55. Functions *g* and *r* are defined by the given graphs.
    a. Determine the values of each of the following
       expressions:
       i. $r(g(2))$ ii. $g(r(1))$
       iii. $r(r(6))$ iv. $g(r(-2))$
    b. How does the output $g(r(x))$ vary as *x* varies from 1
       to 3?

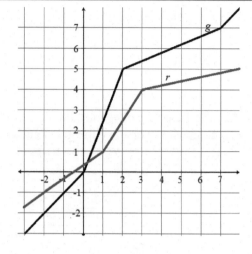

V: EXTRA PRACTICE WITH FUNCTION COMPOSITION (TEXT: S5)

56. The standard formula for determining temperature in degrees Fahrenheit, *F*, when given the
    corresponding measurement in degrees Celsius, *C*, is $F = \frac{9}{5}C + 32$. We can write this formula using
    function notation by letting $F = g(C)$ and writing $g(C) = \frac{9}{5}C + 32$.
    a. State the meaning of *g*(100) and evaluate *g*(100).
    b. Solve the equation *g*(*C*) = 212.
    c. Define a function *h* that converts a temperature measure in degrees Fahrenheit to the
       corresponding measure in degrees Celsius. (*Hint: Solve* $F = \frac{9}{5}C + 32$ *for C.*)
    d. State the meaning of *h*(212) and evaluate *h*(212).
    e. Determine the values of $g(h(212))$ and $h(g(100))$ <u>without performing any calculations</u>. What
       do you notice about the relationship between *g* and *h*?
    f. Represent the value of $g(h(n))$. Represent the value of $h(g(k))$. What do you notice about the
       relationship between *g* and *h*?

57. A spherical bubble inflates so that its volume increases by a constant rate of 120 cm^3 per second.
    a. Define a function *f* that represents the bubble's radius length (in cm), *r*, as a function of the
       bubble's volume (in cubic cm), *V*. (*Note: The volume of a sphere is* $V = \frac{4}{3}\pi r^3$.)
    b. Define a function *g* that expresses the bubble's volume (in cubic centimeters), *V*, as a function
       of the number of seconds since the bubble began to inflate, *t*.
    c. Use your functions from (a) and (b) to define a function *h* that expresses the bubble's radius (in
       centimeters), *r*, as a function of the number of seconds since the bubble began to inflate, *t*.
    d. Use function notation to represent the change in the bubble's radius as the number of seconds
       since the ball began to inflate increases from 5 to 5.3 seconds.

58. A farmer decides to build a fence to enclose a rectangular field in which he will plant a crop. He has
    1000 feet of fencing, and his goal is to maximize the field's area.
    a. Draw a diagram that shows the quantities in this situation. Define variables to represent
       quantities that vary in value.
    b. What are the dimensions of the enclosed field if its width must be 200 feet? What are the
       dimensions of the enclosed field if its length must be 352 feet?
    c. Determine a formula that represents the field's width in terms of the field's length given that the
       total length of the fence is 1000 feet.
    d. Describe how the length of the field changes as the width of the field increases.

e. Determine a formula that relates the length and width of the field to the total area of the enclosed field.

f. Using parts (c) and (e), define a function f that expresses the area of the field (measured in square feet) as a function of the length of the field l (measured in feet).

g. Create a graph that represents the area of the enclosed field in terms of the field's length. Describe how the area of the enclosed field changes as the length of the field increases.

h. Based on your graph, what is the maximum area of the enclosed field? What are the dimensions of the enclosed field that create this maximum area?

59. Janet works 36 hours a week between two part-time jobs during the semester: residence hall front desk assistant ($11.50 per hour) and supermarket cashier ($13.25 per hour).

a. Let r represent the number of hours Janet works in a given week as a residence hall front desk assistant. Write a formula to represent the value of s, the number of hours Janet works as a supermarket cashier in the same week, in terms of r.

b. Define a function f that represents the amount of money Janet makes (in dollars) for working r hours as a residence hall front desk assistant, and then define a function g that represents the amount of money Janet makes (in dollars) for working s hours as a supermarket cashier.

c. Define a function g that represents the amount of money Janet makes as a supermarket cashier in a week while working *r hours as a residence hall front desk assistant that week.*

d. Define a function T that calculates the total amount of money Janet makes in a given week if she is working both jobs and works r hours at the residence hall front desk that week.

e. The supermarket requires Janet to work between 15 and 30 hours each week. What is the domain of function T? What is the range of function T?

60. You are planning a trip to Japan where the currency is the yen. On your way to Japan, you will stop in Italy where the currency is the euro. You know that you will need to convert your U.S. dollars to euros before your trip and know that the number of euros is 0.78 times as large the number of dollars. You also know the number of yen is 103 times as large as number of U.S. dollars.

a. Define a formula that expresses the number of yen, y, you can get by converting d dollars to yen.

b. Define a formula that expresses the number of euros, u, you will get by converting d dollars to euros.

c. You know that you will not be converting directly from dollars to yen because you will first stop in Italy. So, it would be more helpful for you to know the conversion rate between euros and yen. Write a function g that expresses the number of yen you will have, y, in terms of the number of euros you convert, u.

61. The length of a steel bar changes as the temperature changes. Consider a 10-meter steel bar that is placed outside in the morning when the temperature is 20 degrees Celsius. Let l represent the length of the steel bar in meters. Let t be the temperature of the steel bar in degrees Celsius. Let n be the number of minutes since you placed the bar outside. You are given the following relationships: $l = 0.00013(t - 20) + 10$ and $n = 12t - 240$. Write a function f that gives the length of the bar (in meters), l, in terms of the number of minutes elapsed since the bar was placed outside, n.

62. Let $n = 240s + 123$ and let $t = 0.14s^2$. Write a function h that gives n in terms of t.

63. Let $160 = 2t + 4m$ and let $p = \frac{1}{4}tm^2$. Write a function k that gives p in terms of t.

VI. INVERSE FUNCTIONS: REVERSING THE PROCESS (TEXT: S6)

In Exercises #64-66, do the following.
  a. Describe the function process (for example, "the input is increased by 2 and then tripled").
  b. Describe the process that undoes the function (for example, if f is a process that increases its input by 7, the process that undoes f will decrease its input by 7).
  c. Algebraically define a process that undoes the process of the given function – that is, define the inverse function for each of the given functions.

64. $f(x) = \frac{7}{5}x$ 　　　　　　　　　　　65. $g(x) = 10 + x$ 　　　　　　　　　66. $h(x) = \dfrac{x}{12}$

67. a. Algebraically define a process f that multiplies its input by 3, and then decreases this result by 1.
  b. Describe the process that undoes the process of f described in part (a).
  c. Algebraically define the process f^{-1} that undoes the process of f described in part (a).

68. Given the function f defined by $f(x) = 3x + 2$, complete the following.
  a. Define g, the function that undoes the process of f.
  b. Show that $f(g(x)) = x$ using the functions defined in this exercise.
  c. Show that $g(f(x)) = x$ using the functions defined in this exercise.
  d. What do you conclude about the relationship between f and g?

69. Given $g(x) = 14x - 9$, what are the value(s) of the input x when $g(x) = 109$?

70. Given $h(x) = \frac{1}{6}x^2 + 7$, what are the value(s) of the input x when $h(x) = 31$?

71. Given that $h(x) = 4x$ represents a square's perimeter length in terms of its side length, x (both in inches), complete the following.
  a. Algebraically define h^{-1} and describe its input quantity, output quantity, and process.
  b. Describe the input quantity, function process, and output quantity for $h\left(h^{-1}(x)\right)$.

72. Assume that g and z are functions that have inverses which are also functions. Use the words *input* and *output* to express the meanings of each of the following expressions.
  a. $g^{-1}(21)$ 　　　　b. $z^{-1}(32) = 43$

In Exercises #73-76, find the inverse relation for each given function. Then, determine if the inverse relation that you found is a function. Justify your answer.

73. $g(x) = 2x + 4$ 　　　74. $h(x) = 2x^3 - 6$ 　　　75. $k(x) = \frac{1}{2}x^2 + 12$ 　　　76. $m(x) = \dfrac{2}{x+3}$

77. After Julia had driven for half an hour, she was 155 miles from Denver. After driving 2 hours, she was 260 miles from Denver. Assume that Julia drove at a constant speed for her entire trip. Let f be a function that represents Julia's distance (in miles) from Denver in terms of the number of hours she has driven, t.
  a. Define a formula for function f.
  b. Interpret the meaning of $f^{-1}(500)$ and then find its value.
  c. Determine a rule for f^{-1}.
  d. Construct a graph for $f^{-1}(d)$.

78. The functions *f* and *g* are defined in the given table. Based on this table, complete the following.

x	−3	−2	−1	0	1	2	3
f(*x*)	9	2	6	−4	−5	−8	−9
g(*x*)	3	0	3	2	−3	−1	−5

   a. Is f^{-1} a function? Explain why or why not.   b. Is g^{-1} a function? Explain why or why not.
   c. Evaluate the following expressions:
      i. $f^{-1}(-9)$ ii. $g(3)$ iii. $f\left(g(2)\right)$
      iv. $f^{-1}(2)$ v. $f^{-1}\left(g(3)\right)$ vi. $g\left(f^{-1}(-4)\right)$

79. The graphical representations of functions *g* and *h* are given. Use this information to complete the following.
   a. As *x* increases from 0 to 5, how does *g*(*x*) change? Be specific.
   b. As *x* increases from 0 to 5, how does *h*(*x*) change?
   c. Use the graphs above to evaluate the following:
      i. $g^{-1}(2)$ ii. $g^{-1}\left(h(2)\right)$ iii. $g\left(h^{-1}(4)\right)$

80. Given the functions $f(y) = \frac{3}{2}y + 21$ and $g(x) = 2x + 1$, answer the following questions.
   a. Determine f^{-1} and explain how f^{-1} relates to *f* in terms of the input values and output values of each function.
   b. Evaluate $g^{-1}\left(g(2)\right)$ and $g^{-1}\left(g(3.5)\right)$.
   c. What pattern do you observe about $g^{-1}\left(g(x)\right)$? Explain.
   d. Determine the composite function $f\left(g(x)\right)$ and explain, using input-output language, the meaning of the function $f\left(g(x)\right)$ relative to the functions *f* and *g*.
   e. Considering the description of the function $f\left(g(x)\right)$ given in part (d), describe the inverse of this function. Justify whether $f^{-1}\left(g^{-1}(x)\right)$ or $g^{-1}\left(f^{-1}(x)\right)$ is the proper notation for representing this function's inverse, and then determine the rule of the chosen function.

81. Apply the ideas of function inverse and function composition to answer each of the following.
   a. Define the area of a square (in square cm), *A*, in terms of its perimeter length (in cm), *p*.
   b. Define the diameter of a circle *d* in terms of its circumference *C*.

82. Apply the ideas of function inverse and/or function composition to answer each of the following.
   a. Define a square's perimeter length (in cm), *p*, in terms of its area (in square cm), *A*.
   b. If a circle's radius length is growing at a rate of 8 cm per second, define the circle's area (in square cm), *A*, in terms of the number of seconds since the radius length was 0 cm, *t*.
   c. Define the diameter of a circle *d* in terms of its area *A*.

VII. INTRODUCING THE DIFFERENCE QUOTIENT (TEXT: S7)

For Exercises #83-86 do the following.
   a. Determine the function's average rate of change from $x = 2$ to $x = 5$.
   b. Explain what your answer to part (a) represents.
   c. Write an expression that represents the function's average rate of change over any interval of length h. Simplify your expression if possible.

83. $f(x) = 12x + 6.5$ 84. $f(x) = 97$ 85. $f(x) = 3x^3 - 9$ 86. $f(x) = \dfrac{1}{2x}$

87. Recall that the volume of a box V (measured in cubic inches) as a function of the length of the side of the square cutout x (measured in inches) is given by $f(x) = x(11 - 2x)(8.5 - 2x)$ with $V = f(x)$.
   a. Describe the meaning of each of the following expressions.

        i. $f(x+3)$ ii. $f(x+3) - f(x)$ iii. $\dfrac{f(x+3) - f(x)}{(x+3) - x}$

   b. Evaluate $\dfrac{f(x+3) - f(x)}{(x+3) - x}$ when $x = 0.5$. Describe the value's meaning in this context.

88. $f(x) = x^2 - 6x + 10$ represents the altitude of a U.S. Air Force test plane (in thousands of feet) during a recent test flight as a function of elapsed time (in minutes) since being released from its airborne launcher.
   a. Find the average rate of change of the plane's altitude with respect to time elapsed since being released as x varies from $x = 2$ minutes to $x = 2.1$ minutes. Show your work.
   b. What is the meaning of the *average rate of change* you determined in part (a) in this context?
   c. Explain the meaning of the expression $f(k + 0.1)$.
   d. Explain the meaning of the expression $f(k + 0.1) - f(k)$.
   e. You are given the graph of $y = f(x)$ and a specific value of x, say $x = k$. Show $f(k + 0.1) - f(k)$ on the graph.
   f. What does the expression $\dfrac{f(k + 0.1) - f(k)}{(k + 0.1) - k}$ represent?

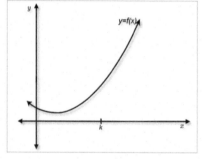

This investigation contains review and practice with important skills and procedures you may need in this module and future modules. Your instructor may assign this investigation as an introduction to the module or may ask you to complete select exercises "just in time" to help you when needed. Alternatively, you can complete these exercises on your own to help review important skills.

Meaning of Exponents
Use this section prior to the module or with/after Investigation 1.

Exponents are used to represent the number of identical factors in a product. For example, in the product $5 \cdot 5 \cdot 7 \cdot 7 \cdot 7$, the factor 5 appears twice and the factor 7 appears three times. Thus, $5 \cdot 5 \cdot 7 \cdot 7 \cdot 7$ is equivalent to $5^2 \cdot 7^2$. The given table shows additional examples. Note that each column represents the same value in three different forms (product notation, exponential notation, and decimal notation).

product notation	$4 \cdot 4 \cdot 4 \cdot 4 \cdot 4$	$2 \cdot 2 \cdot 2 \cdot 2 \cdot 5 \cdot 5 \cdot 5$	$7 \cdot 6 \cdot 4 \cdot 6 \cdot 7$
exponential notation	4^5	$2^4 \cdot 5^3$	$4 \cdot 6^2 \cdot 7^2$
decimal notation	1,024	2,000	7,056

In Exercises #1-3, rewrite each product in exponential and decimal notation.

1. $2 \cdot 6 \cdot 7 \cdot 7 \cdot 6 \cdot 6$

2. $4 \cdot 5 \cdot 5 \cdot 4 \cdot 4 \cdot 4 \cdot 4$

3. $5 \cdot 5 \cdot 5 \cdot 5 \cdot 5 \cdot 3 \cdot 7$

It's important to realize that exponents do <u>not</u> have to be whole numbers. It is possible to evaluate a^b where b is not a whole number. **In Exercises #4-6, use a calculator to evaluate each expression.** Round your answers to three decimal places.

4. $8^{4.3}$

5. $256^{0.5}$

6. $80^{1.715}$

Even if you don't have a calculator, you should be able to estimate the value of exponential expressions with non-whole number exponents. For example, $2^{4.7}$ must be between 16 and 32 because $2^4 = 16$ and $2^5 = 32$. **In Exercises #7-10, estimate the value of each exponential expression by giving two whole number values between which the value must fall.**

7. $3^{3.4}$

8. $10^{2.5}$

9. $7^{0.8}$

10. $4^{-1.9}$

The same number can be represented in terms of ANY base we choose. For example, consider the number 64. Each of the following exponential expressions evaluates to 64. *Take a moment to verify this.*

$$2^6 \quad 4^3 \quad 8^2 \quad 64^1 \quad 4{,}096^{1/2} \quad 16{,}777{,}216^{1/4}$$

However, the following expressions ALSO evaluate to 64 (up to a small rounding error). *Take a moment to verify this.*

$$3^{3.78558} \quad 5^{2.58406} \quad 10^{1.80618} \quad 15^{1.53575} \quad 28^{1.24809} \quad 100^{0.90309}$$

11. Which of the following exponential expressions also represent a value of 64 (up to a small rounding error)? Select all that apply.

   a. $7^{2.137}$

   b. $11^{1.734}$

   c. $9^{1.651}$

   d. $\left(\frac{1}{3}\right)^{-3.786}$

12. Group the following exponential expressions based on their values (up to a small rounding error).
    a. $3^{3.72683}$ b. $12^{1.50415}$ c. $46^{1.02178}$ d. $15^{1.44459}$ e. $10^{1.77815}$ f. $50^{0.95543}$ g. $80^{0.89274}$ h. $215^{0.76236}$

Properties of Exponents
Use this section prior to the module or with/after Investigation 2.

The following definition box shows the most common exponent properties we use to rewrite expressions.

Exponent Properties
For any real numbers a, b, m, and n, the following properties hold.

Product Rule: $a^m \cdot a^n = a^{m+n}$ **Quotient Rule:** $\dfrac{a^m}{a^n} = a^{m-n}$ (if $a \neq 0$)

Power Rule: $(a^m)^n = a^{m \cdot n}$ **Distributive Property:** $(ab)^m = a^m \cdot b^m$

Zero Exponent Rule: $a^0 = 1$ (if $a \neq 0$)   **Negative Exponent Rule:** $a^{-m} = \dfrac{1}{a^m}$ (if $a \neq 0$)

---

13. Write out each of the following in expanded form to demonstrate why the product rule
    $a^m \cdot a^n = a^{m+n}$ is true. *For example, $3^2 \cdot 3^4$ would be written as $3 \cdot 3 \cdot 3 \cdot 3 \cdot 3 \cdot 3$ in expanded form.*
    a. $2^3 \cdot 2^5$ $= 2^8$ b. $4^4 \cdot 4^3$ c. $7^6 \cdot 7^4$

    $(2 \cdot 2 \cdot 2)(2 \cdot 2 \cdot 2 \cdot 2 \cdot 2)$

    2^8

14. Write out each of the following in expanded form to demonstrate why the product rule $(a^m)^n = a^{m \cdot n}$
    is true.
    a. $(2^3)^2$ $= 5^6$ b. $(5^2)^3$ c. $(8^5)^2$

    $(2 \cdot 2 \cdot 2)(2 \cdot 2 \cdot 2)$

    64

15. Write out each of the following in expanded form to demonstrate why the quotient rule $\dfrac{a^m}{a^n} = a^{m-n}$
    is true.
    a. $\dfrac{5^6}{5^2}$ $= 5^4$ b. $\dfrac{3^7}{3^5}$ c. $\dfrac{8^6}{8^5}$

    $\dfrac{5 \cdot 5 \cdot 5 \cdot 5 \cdot 5 \cdot 5}{5 \cdot 5}$

    $\dfrac{3 \cdot 3 \cdot 3 \cdot 3 \cdot 3 \cdot 3 \cdot 3}{3 \cdot 3 \cdot 3 \cdot 3 \cdot 3}$

    9

    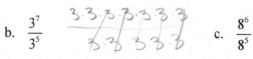
    $5 \cdot 5 \cdot 5 \cdot 5 = 625$

16. Use the quotient rule as well as the expanded form of the following expressions to demonstrate why it makes sense that $a^0 = 1$.

a. $\dfrac{6^4}{6^4}$

b. $\dfrac{2^5}{2^5}$

c. $\dfrac{11^3}{11^3}$

In Exercises #17-24, use the exponent properties to rewrite each of the following using only positive exponents.

17. $\left(x^7\right)^3$

18. $(xy)^4\left(xy^2\right)$

19. $\left(r^2 t\right)^3$

20. $\left(2h^3\right)\left(6h^2\right)$

21. $\left(x^3 y^{-4}\right)\left(x^5 y^2\right)$

22. $\left(a^4 y\right)^2 \left(a^5\right)^0$

23. $\dfrac{14a^5 p^2}{7ap^2}$

24. $\dfrac{x^4 y^{10} z^3}{x^2 y^5 z^6}$

Solving Equations Involving Integer Exponents
Use this section prior to the module or with/after Investigation 4.

Consider an equation like $4x^3 + 9 = 35$. The solution is the value of x such that $f(x) = 4x^3 + 9$ outputs a value of 35.

$$4x^3 + 9 = 35$$
$$4x^3 = 26 \qquad \bullet \text{ subtract 9}$$
$$x^3 = 6.5 \qquad \bullet \text{ divide by 4}$$
$$x = \sqrt[3]{6.5}$$

The solution is $x = \sqrt[3]{6.5}$, or $x \approx 1.86626$.

In Exercises #25-27, solve each equation.

25. $2x^3 - 110 = 140$

26. $2(x^3 + 13) = -28$

27. $\frac{1}{2}x^5 - 20 = -1$

In Exercises #25-27, all of the exponents were odd integers. This is convenient because it created no doubt about the solution. If $x^3 = 64$, then $x = 4$ since $(4)^3 = 64$. If $x^3 = -64$, then $x = -4$ since $(-4)^3 = -64$. If the exponent is even, then this is not the case.

For example, if $x^2 = 25$, then the solutions are $x = 5$ and $x = -5$ because $(5)^2 = 25$ AND $(-5)^2 = 25$. Therefore, if the exponent is even we have to consider that there are two possible solutions to the equation. For example, let's solve the equation $3x^4 - 10 = 7$.

$$3x^4 - 10 = 7$$

$$3x^4 = 17 \qquad \bullet \text{ add } 10$$

$$x^4 = \tfrac{17}{3} \qquad \bullet \text{ divide by } 3$$

$$x = \pm\sqrt[4]{\tfrac{17}{3}}$$

The solutions are both $x = \sqrt[4]{\tfrac{17}{3}}$ AND $x = -\sqrt[4]{\tfrac{17}{3}}$. Note that mathematicians often combine these two solutions into one statement by writing $x = \pm\sqrt[4]{\tfrac{17}{3}}$. It's important to recognize that "$\pm$" is indicating two different values for x.

In Exercises #28-30, solve each equation.

28. $\frac{1}{2}(x^2 + 5) = 82$ 　　　　　　29. $3x^2 - 13 = 41$ 　　　　　　30. $\frac{1}{4}x^6 - 5 = 5$

Solving Equations Involving Variable Exponents (without Logarithms)
Use this section prior to the module or with/after Investigation 5.

The previous two sections were based on solving equations involving power expressions (monomials). But sometimes we have equations where the exponent is unknown. For example, consider the equation $2 \cdot 3^x - 52 = 110$. Its solution is the value of x such that $2(3)^x - 52$ has a value of 110.

We start the solution process as shown.

$$2 \cdot 3^x - 52 = 110$$

$$2 \cdot 3^x = 162 \qquad \bullet \text{ add } 52$$

$$3^x = 81 \qquad \bullet \text{ divide by } 2$$

At this point we have no algebraic method for solving the equation. However, if you know your powers of 3, then you know (or can easily verify) that $3^4 = 81$. So the solution is $x = 4$.

If we instead ended up with something like $5^x = 112$, then for now we can estimate that the solution would be between $x = 2$ and $x = 3$ because $5^2 = 25$ and $5^3 = 125$.

In Exercises #31-33, solve each equation. If the solution is not an integer, then provide the two consecutive integers between which the answer must fall.

31. $3(2^x - 14) = 54$ 　　　　　　32. $\frac{1}{2}(3^x) + 17 = 52$ 　　　　　　33. $6(3^x) - 2 = 28$

Percentage

A ***percentage*** refers to a type of measurement where the measurement unit is a specific proportion of a quantity. To be exact, **1% (of a quantity) is $\frac{1}{100}$ of the quantity's value**.

*1. You likely recall a teacher telling you to "move the decimal point two places to the left" when expressing a percentage as a decimal number. For example, if you want to determine the amount of a 25% tip on a $60 meal for you and a friend, you would compute (0.25)(60). This is because 25% of some quantity's value is 25 (1/100ths) **of the quantity's value** (written in decimal form as 0.25 times as large as the quantity's value).

    a. How do you say "0.25" using proper place value language, and how does this help us understand why 25% (of some quantity's value) and 0.25 (times as large as a quantity's value) are equivalent?

    b. When stating a percentage, such as "35%", we always need to specify the answer to the question, "percentage *of what*?" Explain why this is important and give an example.

    c. Why is the idea of "percentage" useful? Give an example.

    d. Explain what it means to pay 8% sales tax on a purchase of $120. Justify any calculations you perform using the meaning of *percentage*.

---

2. You encounter the idea of "percent (of something)" frequently in your daily life. For each of the following examples,

  (i) perform the calculation,
  (ii) state the units of your answer, and
  (iii) use the definition of percent to explain why your approach makes sense.

  a. Determine the amount of a 15% tip on a food purchase of $85.

  b. Determine the number of hours left on your 9-hour laptop battery that shows 27% of your battery usage remaining. *Assume consistent usage over a 9-hour period.*

  c. If you take out a 1-year loan for $6550 with interest charged at 23% of the amount borrowed, how much money must be repaid at the end of 1 year?

3. Jamie is remodeling his bathroom and is considering the size of square tiles to cover a wall. The side-length of the tile he's considering is displayed by the line segment below. After considering the tile sample, he decided to purchase tile with a side length that is 25% longer than the sample.

sample tile length

  a. Draw a second line above to represent the purchased tile's side length that is 25% longer than the sample's side length. Discuss with a partner or as a class how you determined the length to draw.

  b. Complete the following statements by filling in the blanks, then draw "line diagrams" and label them to demonstrate how to visualize the answer. The first one is done for you as an example.

  i. The purchased tile's side length is ___1.25___ times as long as the sample tile's side length.

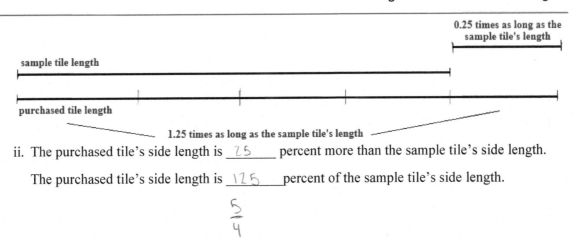

0.25 times as long as the
sample tile's length

sample tile length

purchased tile length

1.25 times as long as the sample tile's length

ii. The purchased tile's side length is __25__ percent more than the sample tile's side length.

The purchased tile's side length is __125__ percent of the sample tile's side length.

$$\frac{5}{4}$$

iii. The sample tile's side length is __0.8__ times as long as the purchased tile's side length.

The sample tile's side length is __80__ percent of the purchased tile's side length.

The sample tile's side length is __20__ percent less than the purchased tile's side length.

$$\frac{4}{5} \text{ or } 80\%$$

$$100\% - 80\% = 20\%$$

Purchased is $\frac{5}{4}$ of sample, sample is $\frac{4}{5}$ of purchased

Percent Change

Percent change refers to the *difference* between two values of a quantity, measured as a percentage of the "starting" value. *Note that a decrease in value is represented with a negative percent change.*

Example 1: If the price of an item was $50 and increased to $59, then we can say the following.
- The change in price is $9.
- The change from the original price to the new price is $\frac{9}{50}$, or $\frac{18}{100}$, times as large as the original price.
- **The percent change from the original price to the new price was 18% of the original price.**
- The new price is $59, which is 100% + 18% = 118% of the original price of $50.

Example 2: If the price of an item was $80 and decreased to $62, then we can say the following.
- The change in price is –$18.
- The change from the original price to the new price is $-\frac{18}{80}$, or $-\frac{22.5}{100}$, times as large as the original price.
- The percent change from the original price to the new price was –22.5% of the original price.
- The new price is $62, which is 100% + (–22.5)% = 77.5% of the original price.

*4. You walk into a store that is having a sale. For each sale described, do the following. [*Draw diagrams to help you.*]
    (i)   State the percent change in the price.
    (ii) State the number we can multiply the original price by to determine the sale price.
    (iii) Find the sale price in dollars.

    a. original price: $150, sale: 40% off   b. original price: $915.99, sale: 15% off

*5. A store needs to raise prices on some of its items. For each price change described, do the following. [*Draw diagrams to help you.*]
    i)   State the percent change in the price.
    ii) State the number we can multiply the original price by to determine the new price.
    iii) Find the new price in dollars.

    a. original price: $52, increase: 10%   b. original price: $13, increase: 50%

6. A new company expanded rapidly and doubled the number of its employees every month over the last year.

    a. Draw a diagram to represent what this means. *Use line segments with lengths representing the number of employees at the company each month.*

    b. *Fill in the blank*: The number of employees in any month is _____% of the number of employees in the previous month.

    c. What was the percent change in the number of employees from one month to the next? How can you represent this in your diagram?

When examining situations with quantities that vary in value, we typically want to identify pairs of quantities and examine how their values change together (co-vary). When we identify such a pair, our next goal is to understand *how* their values change together and, if possible, to identify patterns in how the output value changes when we repeatedly add a fixed amount of change to the input quantity's value. One approach to studying patterns in co-varying relationships is to identify what stays the same as the quantities' values change.

We previously studied **linear functions**. A linear relationship exists when two quantities' values change together with a **constant rate of change**. In such a relationship, the two quantities' values co-vary, but they must do so in such a way that the ratio of their changes, $\frac{\Delta y}{\Delta x}$, is a constant. Not all co-varying relationships exhibit a constant rate of change. In this investigation, we will explore patterns of change for **exponential functions** and identify what remains the same as the input and output quantities co-vary.

For Exercises #1-5, use the following information. When a piece of new technology is invented and sold, its value decreases over time because it wears down and because newer, better technology is developed to replace it. Suppose two electronic devices have the same resale value right now and that their resale values are expected to change according to the following patterns.

- **Product #1: iTech Device** The current resale value is $300. The resale value is expected to decrease 30% per year for the next several years.
- **Product #2: Dynasystems Device** The current resale value is $300. The resale value is expected to decrease $45 per year for the next several years.

*1. a. Find the expected resale value (in dollars) for both devices 1 year from now and demonstrate how you determined each value.

iTech: $300 - (300 \cdot 0.3) = \$210$
Dyna. $300 - 45 = \$255$ } after one year

b. Without performing additional calculations, predict which device will have a greater resale value 6 years from now.

iTech

c. Complete the table showing the expected resale value for each device over the next 6 years.

Δt	Years from now, t	Expected resale value for the iTech Device, s (dollars)	Δs	Δt	Years from now, t	Expected resale value for the Dynasystems Device, n (dollars)	Δn
	0	300			0	300	
+1			-90	+1			-45
	1	210			1	255	
+1			-63	+1			-45
	2	147			2	210	
+1			-44.1	+1			-45
	3	102.9			3	165	
+1			-30.87	+1			-45
	4	72.03			4	120	
+1			-21.609	+1			-45
	5	50.421			5	75	
+1			-15.1263	+1			-45
	6	35.2947			6	30	

d. Check your prediction from part (b) by using the table in part (c).

Correct!

e. The table on the left demonstrates how the expected resale value for the iTech Device co-varies with elapsed time. What stays the same as these two quantities' values change together?

The percent change/decrease is the same each year (always -30%) for both price and devaluation.

f. The table on the right demonstrates how the expected resale value for the Dynasystems Device co-varies with elapsed time. What stays the same as these two quantities' values change together?

The value that the resale price decreases by is the same each year

*2. a. Define function formulas to model the expected resale value in dollars for each device as functions of the number of years from now, t.

iTech Device: $s = f(t)$; where $f(t) =$ $300(0.7^t)$

Dynasystems Device: $n = g(t)$; where $g(t) =$ $300 - 45t$

b. Explain what each part of both formulas represents relative to the context we are exploring.

$f(t) = 300(0.7^t)$

1 $300 \cdot (0.7)$

2 $300 \cdot (0.7 \cdot 0.7)$

3 $300 \cdot (0.7 \cdot 0.7 \cdot 0.7)$

$g(t) = 300 - 45t$

1 $300 - 45$

2 $300 - 45 - 45$

3 $300 - 45 - 45 - 45$

3. Find and interpret the following expressions using your functions from Exercise #2.
   a. $f(4)$

   b. $g(3)$

   c. $g(6) - g(2)$

   d. $f^{-1}(147)$

*4. In Exercise #1, the table on the left has a column showing values of $s = f(t)$ (the expected resale value for the iTech Device in dollars) as well as a column showing values of Δs (the change in the expected resale value for the iTech Device in dollars).

a. What are the ratios of consecutive entries in the column for s? That is, what are the values of $\frac{f(1)}{f(0)}, \frac{f(2)}{f(1)}, \frac{f(3)}{f(2)}$, and so on?

$$\frac{210}{300} \Big)\quad \frac{147}{210} \Big)\quad \frac{102.9}{147} \Big)$$
$$0.7\ell \qquad 0.7\ell \qquad 0.7\ell$$

b. How does your answer to part (a) relate to the original problem context? In other words, what do these ratios tell us about the resale value of the iTech Device?

Each year the resale value is 70% of the resale value of the previous year.

c. What are the ratios of values of Δs compared to the values of s at the beginning of each interval?

$$\frac{-90}{300} = -0.3 \qquad\qquad \frac{-63}{210} = -0.3 \qquad -\frac{44.1}{147} = -0.3$$

d. How does your answer to part (c) relate to the original problem context? In other words, what do these ratios tell us about the resale value of the iTech Device?

The decrease is always 30% of current price

You are given the graphs of functions f and g that model the expected resale values for each device over the next 6 years. On each graph we have plotted the points that correspond to the ordered pairs from the tables in Exercise #1.

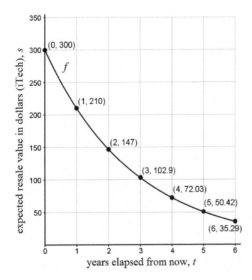 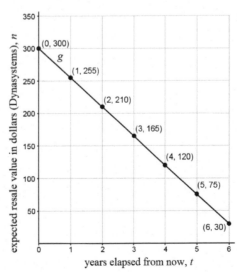

To help focus attention on the values of *s* and *n* (the expected resale values of each device in dollars) represented in our previous tables, we have drawn vertical line segments whose lengths represent values of *s* and *n*, respectively.

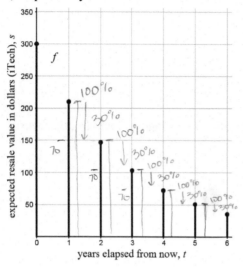 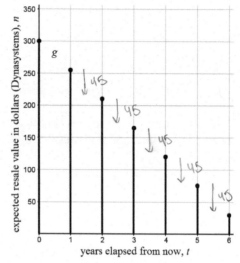

*5. a. For the graph on the left, explain how we can "see" the fact that the expected resale value of the iTech Device at some moment is 70% of its expected resale value one year earlier.

   b. For the graph on the left, explain how we can "see" the fact that the expected resale value of the iTech Device changes by –30% each year.

   c. Explain how we can "see" on the graphs that only one of the two devices has an expected resale value that changes at a constant rate of change with respect to the number of elapsed years.

Exponential Functions

When two quantities change together such that, for equal changes in one quantity's values, the second quantity's values have a ***constant percent change***, then we say that the relationship is an ***exponential function***.

---

*6. In Exercises #1-5, you explored an example of a relationship that can be modeled with an exponential function. Using your work with this example, explain why exponential functions ***do not*** have a constant rate of change of one quantity with respect to the other quantity.

In the previous investigation, we established the meaning of **exponential growth**. In this investigation, we will explore the implications of exponential growth in various situations and formalize the idea of a **growth factor**.

*1. How large do numbers become if you repeatedly double them? Imagine you have a chessboard and place 2 pennies on the first square, double that to 4 pennies on the second square, double that to 8 pennies on the third square, and so on.

   a. There are 64 squares on a chessboard. <u>Without doing any calculations</u>, make two <u>predictions</u>.

   i. About how many pennies do you think will be on the last square? Will it match something like the number of students at your school or university? The population of Los Angeles? The population of the United States? Something else?

   ii. How tall will the stack of pennies be on the 64th square? As tall as a house? As tall as a skyscraper? Would the stack reach from Earth to the Moon? Something else?

   b. Fill in the given table and use it to create a formula and graph relating the number of pennies on a square with the square's position number. Be sure to label and scale the axes.

   Table: Graph: Formula (define your variables):

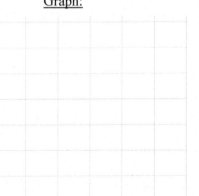

Square Number	Number of Pennies
1	
2	
3	
4	
…	…
10	
…	…
30	

   c. How does the number of pennies on the chessboard vary as you move along the board? For example…

   i. Each time the square position increases by 1, what happens to the number of pennies?

   ii. Each time the square position increases by 2, what happens to the number of pennies?

   iii. Each time the square position increases by 3, what happens to the number of pennies?

   d. What is the *percent change* in the number of pennies on a square when the square position increases by 1?

e. How many pennies will be on square 64? How does this compare to your estimate in part (a)?

2. The function $f(x) = 2^x$ is an example of an exponential function, and some people call this a *doubling function*. The function in Exercise #1 was a doubling function, but its domain was restricted to positive integers. We can also think about continuous doubling functions.

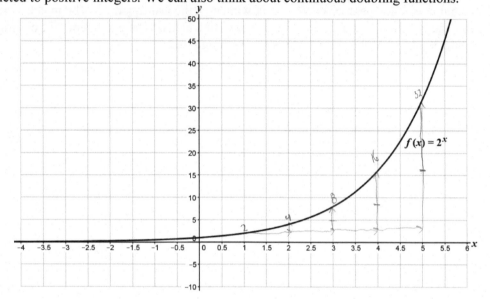

a. Determine the value of $f(0)$.

$$f(0) = 1$$

b. i. When x increases by 1, the new output value is __200__ % of the previous output value.

    ii. What is the percent change in the output value when x increases by 1? 100%

    It is 100% MORE than the previous output value

c. As x increases away from $x = 0$, how do the function values change? What happens as x gets very, very large?

d. As x decreases away from $x = 0$, how do the function values change? What happens as x gets very, very small (becomes a large magnitude negative number)?

e. Starting from the horizontal axis, draw perpendicular vertical line segments that end at $(1, f(1))$, $(2, f(2))$, $(3, f(3))$, $(4, f(4))$, and $(5, f(5))$. Discuss as a class how these line segments help you visualize the relationship explored in this exercise.

3. Let $g(x) = 3^x$ and $h(x) = 6\left(\frac{2}{3}\right)^x$.

   a. Add these functions to the graph in Exercise #2.

   b. With a partner or as a class, discuss how these functions compare to each other and to $f(x) = 2^x$. In particular, discuss the following questions.

      i. What is the vertical intercept for each function?

      ii. As x increases, how do the function values change? What happens as x gets very large?

      iii. As x decreases, how do the function values change? What happens as x gets very small (becomes a large magnitude negative number)?

      iv. What is the 1-unit percent change in the function values?

100% + % change = growth factor

(1-Unit) Growth Factor

For an exponential function, the ratio of output values is always the same for equal-sized intervals of the domain. When x increases from x_1 to x_2, this ratio $\frac{f(x_2)}{f(x_1)}$ is called a **growth factor**. If the interval is 1 unit wide (that is, if the change in x from x_1 to x_2 is 1), then the ratio is the **1-unit growth factor**. 1 year → 2 year → 3 years

A growth factor is a useful number because it tells us how many times as large $f(x_2)$ is compared to $f(x_1)$, and thus it can be used to determine $f(x_2)$ if we know $f(x_1)$.

Note that if this factor is between 0 and 1, it is often called a **decay factor** because a multiplier between 0 and 1 means that the function values are decreasing as x increases.

---

4. From about 2020 to 2025, home prices in the United States are projected to increase dramatically. From 2020 to 2025, the value of a specific house will increase by 16% per year.

   a. Without performing any calculations, predict the home's value after 5 years if its value increases 16% per year. (Will the value be 50% larger? Double the original value? Something else?)

   b. Complete the following table and draw a graph modeling the value of this house in dollars *n* years since the beginning of 2020. Be sure to label the axes.

years since the beginning of 2020	home price (in dollars)
0	280,000
1	324,800
2	376,768
3	437,050.88
4	506,979.0208
5	588,095.66413

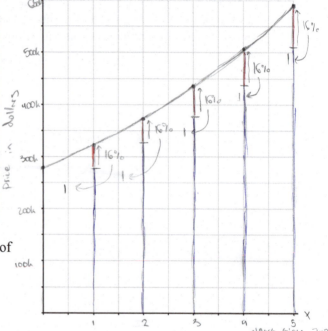

   c. By what number can we multiply the value of the home at one moment in time to find its value one year later?

   1.16

   d. Fill in the blank: The value of the home at one moment in time is __116__ % of its value one year earlier.

   e. Explain how we can "see" the value of the multiplier you identified in part (c) in the graph from part (b).

   ● = 100%   ○ = 16% of

   f. The home's value at the beginning of 2025 was how many times as large as its value at the beginning of 2020? How does this compare to your prediction in part (a)?

   g. What does your model predict the home's value will be at the beginning of 2050? Do you think this number is reasonable?

*5. Some of your classmates made the following claim. "The functions $f(x) = x^2$ and $g(x) = x^3$ are some other examples of exponential functions."

    a. Are your classmates correct? Justify your answer using what you've learned so far in this module.

    b. If you agree with your classmates, then come up with at least two more examples of exponential functions. If you disagree with your classmates, then give a possible reason for why this might be a common mistake for students.

*6. Let $g(p) = 1.578(0.68)^p$.

    a. What does the number "1.578" represent for this function?

    b. What does the number "0.68" represent for this function?

    c. *Fill in the blank*: Whenever p increases by 1, the new output value is _____% of the old output value.

    d. What is the 1-unit percent change?

7. After having 1.4 million people at the start of 2020, the population of a city has been decreasing by 2.1% per year.

    a. What is the 1-year percent change in the city's population?

    b. *Fill in the blank*: When the time elapsed since the beginning of 2020 increases by 1 year, the new population is _____% of the old population.

    c. What is the 1-year growth or decay factor and what does this value tell us about the situation?

    d. Write a function formula to model the city's population (in millions) in terms of the time elapsed since the beginning of 2020 (in years).

In Exercises #8-11, do the following.
   a. Find the ratio of output values that corresponds to increases of 1 in the input value (this value is the 1-unit growth/decay factor).
   b. Determine the 1-unit percent change by comparing the change in the output values to the function value at the beginning of a 1-unit interval for x.
   c. Identify or determine the value of the function when $x = 0$.
   d. Use the information from parts (a) through (c) to define a function formula for the relationship.

8.

x	0	1	2	3
$f(x)$	16	4	1	0.25

a.

0.25

b.

-75%

c.

16

d.

$f(x) = 16(0.25^x)$

*9.

x	1	2	3	8
$g(x)$	260	299	343.85	691.605

a.

b.

c.

d.

10.

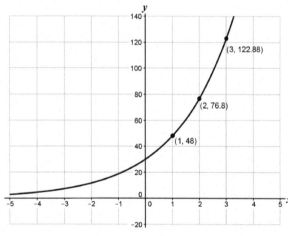

a.

b.

c.

d.

*11.

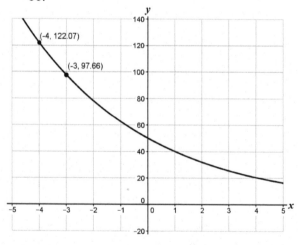

a.

b.

c.

d.

*1. Express your understanding of exponents by answering the following questions.

    a. i. What does b^5 represent?

       ii. Evaluate 4^3 and say what your answer represents.

    b. Calculate the following: i. $9^{1/2}$ ii. $4^{1/2}$ iii. $121^{1/2}$

    c. In general, how do you go about determining the value of some number b raised to an exponent of ½ and why?

    d. In general, how do you go about determining the value of $x^{1/3}$?

    e. Calculate $4^{3/2}$ (also represented as $4^{1.5}$) using the properties of exponents.

    f. Solve the following equations for x:

      i. $x^2 = 81$ ii. $x^3 = 27$ iii. $x^{3/4} = 8$ iv. $x^{1/5} = 2$

    g. Simplify each of the following.

      i. $a^2 \cdot a^4$ ii. $\dfrac{b^7}{b^2}$ iii. $(c^2)^5$

*2. You are given that (4, 38) and (7, 50) are two ordered pairs for an *exponential growth function*. Your classmate noticed that the difference in the outputs is 12 as the inputs change by 3, so they created the given table showing some additional ordered pairs.

x	y
4	38
5	**42**
6	46
7	**50**

Explain how you know that the two additional ordered pairs (bolded) cannot be accurate even though we don't have the function formula or the graph for this relationship.

[handwritten] There isn't exponential growth, each time x increases by 1 y increases by 4.

[handwritten, pointing to x column] this is a constant rate of change, linear

3. From March 16 to March 20, 2024, the total number of cases of the flu in a school tripled every 2 days, following a pattern of exponential growth.
   a. What was the 1-day growth factor and 1-day percent change for the total number of flu cases in this school during this period?

   b. What was the 4-day growth factor and 4-day percent change for the total number of flu cases in this school during this period?

   c. There were 975 total cases of the flu in this school on March 16, 2024. Assuming the pattern of growth continued, give two methods (based on your work in parts (a) and (b)) to determine the total number of cases of the flu in this school on March 24, 2024.

*4. A farm in Canada had 218 alpacas on January 1, 2016. After 8 years, the alpaca population had decreased to 187. Assume the number of alpacas decays exponentially.
   a. What is the 8-year decay factor?

   [handwritten] 0.8577981651

b. Fill in the blank: The number of alpacas on the farm on January 1, 2024 was ___85.78___ % of the number of alpacas on the farm on January 1, 2016.

c. What is the 8-year percent change in the alpaca population?

-14.22%

d. Assuming the alpaca population continues to be modeled by the same exponential model, how many alpacas can we expect to be on the farm on January 1, 2032?

~ 160 alpacas

e. Since 8-year changes in time are quite long, we might want to know how the alpaca population changes over shorter periods of time. What is the 1-year decay factor in the alpaca population?

0.9810095

f. What is the 1-year percent change in the alpaca population?

-1.899%

g. Define a function *f* that models the number of alpacas on the Canadian farm *t* years from January 1, 2016. (*Assume the alpaca population continues to change by the same decay factor each year.*)

$f(t) = 218(0.8578^t)$

*5. *Continue with the same context from Exercise #4*. Assume the number of alpacas continues to change by the same decay factor each year. Use your calculator to complete the following.
  a. How many years after January 1, 2016 will there be 55 alpacas?

  b. After how many years will there be 109 alpacas remaining on the farm, assuming the herd started with 218 alpacas?

  c. At some point, the herd will contain 120 alpacas. From that time, how much longer will it take for the herd to decrease to 60 alpacas?

d. What is the 1-month decay factor in the alpaca population?

0.9984035

e. Define a function g that models the number of alpacas on the Canadian farm k months since January 1, 2016. (*Assume the alpaca population continues to change by the same decay factor each month.*)

6. After taking medicine, your body begins to break it down and remove it according to a pattern of exponential decay. Suppose you take 500 mg of Ibuprofen and, after 5 hours, only 110 mg remain in your bloodstream.
   a. After 5 hours, what percentage of the 500 mg dose remains in your bloodstream?

   b. Define a function f that expresses the amount of Ibuprofen in milligrams (mg) present in your bloodstream, B, after n 5-hour time intervals since taking the medicine.

   c. Define a function g that expresses the amount of Ibuprofen in milligrams (mg) present in your bloodstream, B, t hours after taking the medicine.

   d. How much Ibuprofen remains in your body 3 hours after taking the medicine? After 12 hours?

   e. How does the *change* in the amount of Ibuprofen remaining change over successive 1-hour intervals as the number of hours elapsed increases?

For Exercises #1-2, determine the specified growth or decay factors, percent change, and initial value for each of the following exponential functions.

*1. $f(x) = 9.5(1.24)^x$

    a. 1/2-unit Growth Factor:

$$\sqrt{1.24} = 1.11355$$

    b. 2-unit Growth Factor:

$$1.24^2 = 1.5376$$

    c. 2-unit Percent Change:

$$53.76\%$$

    d. Initial Value:

$$9.5$$

*2. $g(x) = 0.46(0.874)^{4x}$

    a. 1/4-unit Decay Factor:

    b. 1-unit Decay Factor:

    c. 5-unit percent change:

    d. Initial Value:

In Exercises #3-4, determine the specified growth or decay factors, percent change, initial value, and function formula for each of the tables modeled by exponential functions.

3.

x	0	2	4	6
$f(x)$	16	10.24	6.554	4.194

    a. 2-unit Decay Factor:

$$10.24 / 16 = 0.64$$

    b. 1-unit Percent Change:

 $\sqrt{0.64} = 0.8$

$$-20\%$$

    c. 3-unit Decay Factor:

$$\sqrt{0.64} = 0.8^3 = 0.512$$

    d. ½-unit Decay Factor:

$$\sqrt{0.8} = 0.894427191$$

    e. Initial Value: 16

    f. Formula: $f(x) = 16(0.8^x)$

*4.

x	1	4	7	10
$g(x)$	260	278.2	297.674	318.511

    a. 3-unit Growth Factor:

$$278.2 / 260 = 1.07$$

    b. 6-unit Percent Change:

$$1.07^2 = 1.1449 = 14.49\%$$

    c. 1-unit Growth Factor:

$$\sqrt[3]{1.07} = 1.0228091218$$

    d. ¼-unit Growth Factor:

$$\sqrt[4]{1.0228091218} = 1.005654454$$

    e. Initial Value: $260 / 1.0228091218 = 254.20188$

    f. Formula: $g(x) = 254.20188(1.0228^x)$

(margin note, center) multiply by GF → divide by ← GF

In Exercises #5-6, determine the specified growth or decay factors, percent change, initial value, average rate of change, and function formula for the exponential function based on the given graph.

*5. Use the graph to complete the following.

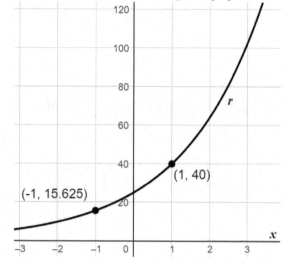

(-1, 15.625)

(1, 40)

a. 2-unit Growth/Decay Factor:

$40 / 15.625 = 2.56$

b. 6-unit Growth/Decay Factor:

$2.56^3 = 16.777216$

c. 1-unit Percent Change:

$\sqrt{2.56} = 1.6$

60%

d. ½-unit Growth/Decay Factor:

$\sqrt{1.6} = 1.264911064 1$

e. Initial Value:

$15.625 \cdot 1.6 = 25$

f. Function Formula:

$r(x) = 25(1.6)^x$

g. The average rate of change of $r(x)$ with respect to x as the value of x increases from -1 to 1.

h. Explain how to interpret the value you found in (g).

*6. Use the graph to determine the following values.
   a.   3-unit Growth/Decay Factor:

   $36.84160 = 0.614$

   b.   6-unit Growth/Decay Factor:

   $0.614^2 = 0.376996$

   c.   1-unit Percent Change:

   -15.006%

   d.   ½-unit Growth/Decay Factor:

   0.92192316686

   e.   Initial Value:

   60

   f.   Function Formula:

   $w(x) = 60(0.84994^x)$

   g.   The average rate of change of $w(x)$ with respect to x as the value of x increases from 0 to 3.

   h.   Explain how to interpret the value you found in (g).

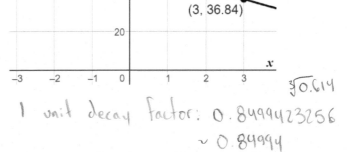

$\sqrt[3]{0.614}$

1 unit decay factor: 0.8499423256

~ 0.84994

$1 - 0.84994 = 0.15006$

In Exercises #7-8, determine the growth or decay factor, percent change, initial value, and function formula for each exponential situation.

*7. From the start of 2006 to the start of 2011, the number of people in the U.S. living in poverty increased exponentially from 36.46 million people to 46.34 million people.

    a. 5-year Growth Factor:

$$46.34 / 36.46 =$$
$$1.270981898$$

    b. 1-year Percent Change:

$$\sqrt[5]{1.270981898} = 1.0491265386$$
$$4.912\%$$

    c. Initial Value:

    d. Function Formula:

$$f(x) = 36.46(1.04913)^x \longleftrightarrow 36.46$$

$$x = \text{years since } 2006$$

8. After having 1.97 million people in 2010, the population of a city has been decreasing by 4.2% every three years.

    a. 1-year Decay Factor:

    b. ¼-year Percent Change:

    c. Initial Value:

    d. Function Formula:

*9. The amount of caffeine in your body decreases by 27% every 4 hours.

    a. 24-hour Decay Factor:

    b. 24-hour Percent Change:

    c. 1-hour Decay Factor:

    d. 1-hour Percent Change:

When you deposit money in a bank, the bank pays you interest based on the balance (the amount of money in the account). The amount of interest paid is some percentage of your account balance, making the amount of interest paid proportional to the amount of money in your account. As a simple example, a bank that advertises an interest rate of 6% per year (compounded once per year) pays each customer 0.06 times as much as their account balance once each year. The new balance after one year (assuming no money is added to or taken from the balance) is then 1.06 times as large as the balance at the beginning of the year.

Interest can be paid in different ways, such as once a year, twice a year, once a month, etc. By convention, banks calculate the interest rate per compounding period by dividing the annual interest rate by the number of compounding periods in one year. ***Note that this technique is not the same as determining partial growth factors.***

*1. You decide to invest $1000 and have three choices of investment accounts. Each account advertises an interest rate of 8%.

a. The first account advertises, "an interest rate of 8% compounded annually." At the end of each year, 8% of the current balance is added to the value of the account.

i. What is the interest rate per compounding period?

ii. Complete the given table.

iii. What is the annual growth factor for this account? Interpret the annual growth factor in the context of this problem.

8% Interest Rate Compounded Annually	
number of years since the investment was made	value of the investment (in dollars)
0	
1	
2	
t	

b. The second account advertises "an interest rate of 8% compounded semiannually." At the end of every 6 months, interest is added to the account balance.

i. What is the interest rate per compounding period?

8% Interest Rate Compounded Semiannually (Every 6 Months)	
number of years since the investment was made	value of the investment (in dollars)
0	
0.5	
1	
2	
t	

ii. Complete the given table.

iii. What is the six-month growth factor for this account?

iv. What is the annual growth factor for this account? Interpret the annual growth factor in the context of this problem.

v. Use the annual growth factor computed in part (iv) to determine the actual percentage change of the account over a 1-year period. This percentage is called **the annual percentage yield (APY).**

c. The third account advertises "an interest rate of 8% compounded daily." 365 times per year (that is, every day), interest is added to the account's value.

i. What is the interest rate compounding period?

365

8% Interest Rate Compounded Daily		per
number of years since the investment was made	value of the investment (in dollars)	
0	1000	
$\frac{1}{365}$	1000.21918	
$\frac{2}{365}$	1000.4384	
1	1083.2776	
2	1173.4903	factor
t	$f(t) = 1000(1.0833)^t$	

ii. Complete the given table.

iii. What is the daily growth for this account?

$8/365 = 1.000219$

iv. What is the annual growth factor for this account? Interpret the annual growth factor in the context of this problem.

1.0833

v. Use the annual growth factor computed in part (iv) to determine the annual percentage yield (APY).

8.33%

d. Which of the three accounts would you choose for your investment? Justify your answer.

Compound Interest Formula

The value of an account (in dollars), A, is represented by the following function f.

$$f(t) = P\left[\left(1+\tfrac{r}{n}\right)^n\right]^t$$

$$= P\left(1+\tfrac{r}{n}\right)^{nt}$$

where

- $A = f(t)$,
- t represents the number of years since the account was opened,
- P represents the initial account value (the account value when $t = 0$),
- n represents the number of times interest is compounded (how many times interest is calculated and deposited into the account) each year, and
- r represents the interest rate expressed as a decimal instead of a percentage.

2. Explain what quantity's value is represented by each of the following expressions.

a. $1+\tfrac{r}{n}$

1 - comp period growth factor
(day, week, month)

b. $\left(1+\tfrac{r}{n}\right)^n$

One year GF

$\tfrac{r}{n} = 1 -$ comp period percent change

c. $\left[\left(1+\tfrac{r}{n}\right)^n\right]^t$

t-year growth factor

d. $P\left[\left(1+\tfrac{r}{n}\right)^n\right]^t$

Total value of account in dollars

3. Bank USA is offering a 9% interest rate compounded monthly, while Southwest Investment Bank is offering a 8.5% interest rate compounded daily. Which bank would you choose for your investment?

---

*4. You should have noticed that the techniques in this investigation for finding growth factors related to different units of time are not the same as the techniques used in previous investigations. This is because bank policy differs from our goals in creating partial growth factors. Look at the following comparison.

<table>
<tr><td align="center"><u>**Non-Financial Contexts**</u></td><td align="center"><u>**Compound Interest**</u></td></tr>
<tr><td>

The population of a certain country is currently 26.4 million people and is increasing by 4% per year.
Annual growth factor: 1.04
Annual percent change: 4%
Monthly growth factor:
$$(1.04)^{1/12} \approx 1.003274$$
Monthly percent change: 0.3274%

When the monthly growth factor is applied 12 times, it produces the annual growth factor.
$$[(1.04)^{1/12}]^{12} = 1.04$$

</td><td>

$1200 is deposited into an account with an interest rate of 4%.
If interest is compounded once per year, the annual growth factor is: 1.04
The annual percent change is: 4%
If interest is compounded once per month, the monthly growth factor is: $1 + \frac{0.04}{12} = 1.00\overline{3}$

The monthly percent change is: $0.\overline{3}\%$

If interest is compounded once per month over a period of one year, the annual growth factor is different than when interest is compounded only once per year.
$$(1.00\overline{3})^{12} \approx 1.04074$$

</td></tr>
</table>

a. In non-financial contexts, we know the actual annual growth factor based on data, and we cannot change the long-term growth pattern of the function relationship. How does this play a role in the methods we use for computing growth factors for different-sized changes in the input quantity's value?

b. Why don't banks have the same restrictions as those described in part (a), and how might this impact their techniques?

*5. A deposit of $5,000 is made into an account paying an interest rate of 5%. Determine the amount in the account 10 years after the investment was made if the interest is compounded:

a. Annually

b. Monthly

c. Weekly

d. Daily

*1. $1000 is invested into an account with an interest rate of 8%. Given the following compounding periods, determine the account's value at the end of the fifth year. Also, determine the annual percent yield (APY) for each compounding period. (*Round to at least 3 decimal digits.*)

	APY (%)	Difference in the value of the APY (%)	Value of the Investment after 5 years	Difference in the value of the investment
Compounded Yearly	8%		$1,469.32808	
		+0.243216%		+$16.61932
Compounded Quarterly	8.243216%		$1,485.9474	
		+0.057%		+$3.90
Compounded Monthly	8.300%		$1489.85	
		+0.028%		+$1.91
Compounded Daily	8.328%		$1491.76	
		+0.00091%		+$0.06
Compounded Hourly	8.329%		$1491.822	
		+0.0000398%		+$0.003
Compounded Every Minute	8.329%		$1491.825	

2. How is the number of compounding periods per year related to the annual growth factor and the APY? Why do you think this is happening?

   As the number of compounding periods increases, APY and annual growth increases.

*3. a. What do you anticipate the APY will be if the interest rate is 8% and the $1000 investment is compounded every second? Every 1/100th of a second?

   Close to 8.329%

   b. What do you predict is the largest annual growth factor possible given an interest rate of 8%? Interpret the meaning of this value in the context of this situation.

   Close to 1.08329

*4. Consider investment interest rates of 5%, 6%, 7%, and 8%.
   a. Complete the first row of the given table by increasing the number of compounding periods per year to 10,000 and predicting the largest annual growth factor that corresponds to each interest rate. (*Round to at least 5 decimal digits.*) Do not complete the second row yet.

Interest Rate	5%	6%	7%	8%
Largest Annual Growth Factor, g	1.051271093	1.06184	1.07251	1.08329
e^r (where r is the interest rate as a decimal)	$e^{0.05}$ 1.05271			

b. e is a constant irrational number (like π) with a value of approximately 2.71828. Its value can be seen by examining the value of the expression $\left(1+\frac{1}{n}\right)^n$ as n increases without bound. Take a moment to do this.

c. Complete the second row of the table. What do you notice? What does this tell us?

An interesting mathematical idea is that no matter what the interest rate is for a given investment, the largest annual growth factor for that investment is related to the number 2.71828... This irrational number is referred to as e.

> **_e_**
>
> e is a constant irrational number with a value approximately equal to 2.718281828459...

*5. An investment of $1000 was made in four different banks at the given interest rates.

Interest Rate	5%	6%	7%	8%
Largest Annual Growth Factor (decimal form)				
Exact Representation of the Largest Annual Growth Factor				
Maximum Value of Investment After t Years				

a. Complete the first row in the table by finding the largest annual growth factor for each interest rate as a decimal approximation.

b. Complete the second row in the table by writing an expression that determines the **exact** value of the largest possible annual growth factor for each interest rate.

c. Complete the third row in the table by writing an expression that can be used to find the maximum value of the investment after t years (the value of the investment assuming the largest possible annual growth factor).

*6. Define a function f that represents the maximum value of an investment (in dollars), B, in terms of the number of years since the initial investment was made, t. Let P represent the value of the initial investment (in dollars) and let r represent the decimal form of the interest rate.

Another way to think about the largest possible annual growth factor for accounts earning compounded interest is to say that it is the growth factor that corresponds to compounding as many times as possible in a year. In other words, if the investment were to be compounded more often than daily, more often than hourly, more often than every second, etc., then we would say the investment is being ***compounded continuously***.

7. $1500 is invested into an account with an advertised interest rate of 7% compounded continuously.
   a. Determine the value of the investment after 6 years.

   b. Determine the amount of time required for the value of the investment to double.

   c. Determine the value of the account after 6 years had the investment been compounded quarterly instead of continuously.

Continuously Compounded Interest Formula

The value of an account (in dollars), B, is represented by the following function f **if interest is compounded continuously**.

$$f(t) = P\left[e^r\right]^t$$
$$= Pe^{rt}$$

where

- $B = f(t)$,
- t represents the number of years since the account was opened,
- P represents the initial account value (the account value when $t = 0$),
- r represents the interest rate expressed as a decimal instead of a percentage.

Note that e^r thus represents the annual growth factor for the account's value.

Up to this point, we have been unable to algebraically solve equations involving exponential growth—equations in which we need to solve for the input variable in the exponent given a value of the dependent variable. We utilized technology to solve these equations by creating graphs. In this investigation, we will learn how to solve these equations algebraically.

*1. a. Without a calculator, determine or approximate the solution to each equation. (*Think about what value of x makes each equation true.*)

    i. $2^x = 16$ ii. $2^x = 10$ iii. $3^x = 1/81$

    b. Describe the process (thinking) you used to determine your solutions in part (a).

Determining the equations' solutions in Exercise #1 involved undoing exponentiation. Instead of raising a base to an exponent (or specifying a number of factors of the base to be multiplied together), we determined the exponent (the number of factors) of the base that results in some number (a product).

As seen in Exercise #1, it can be difficult to determine the exact number of times a base value is a factor of some other number. The standard way of expressing the specific value of an unknown exponent, when provided with both the base value and the result of exponentiation, utilizes what we call *logarithmic notation*. As an example, to represent the number of times 5 is a factor of 125, we write $\log_5(125)$. In general, we say, $\log_b(m)$ represents the number of times b is a factor of m.

Logarithms

For $b > 0$, $b \neq 1$, and $m > 0$, $\log_b(m)$ represents the number of times b is a factor of m.

how many factors of b are in the result (product) m how many factors of 5 are in the result (product) 125

$$\log_b(m) \qquad\qquad \log_5(125)$$

base (factor)   result (product)   base (factor)   result (product)

---

*2. Using logarithmic notation, *represent* the exact solution to each equation.

    a. $2^x = 16$ (4)   b. $2^x = 10$ c. $3^x = 1/81$ (-4)

    $\log_2(16) = x$ $\log_2(10) = x$ $\log_3(1/81) = x$

The function that undoes the process of exponentiation is called the *logarithmic function*. The input of the logarithmic function is the output of the exponential function (the result of raising some base to an exponent), and the output of the logarithmic function is the input of the exponential function (the exponent to which the base is raised). We use "log" as an abbreviation for "logarithm" in expressions.

As an example, $\log_2(10) = x$ is read "log base 2 of 10 equals x" and is equivalent to $2^x = 10$ in exponential form. The number 10 represents the result of raising 2 to an exponent x. The value of x can be determined by answering the question, "To what exponent do we raise the base 2 to obtain the value 10?"

> ### Logarithmic Function
>
> For $x > 0$ and $b > 0$, with $b \neq 1$, the statement $y = \log_b x$ is equivalent to $b^y = x$. The function $f(x) = \log_b(x)$ is the logarithmic function with base b.
>
> We can rewrite any exponential equation in logarithmic form. *Note that the input to the logarithmic function can only be positive real numbers.*

*3. Rewrite each exponential equation in logarithmic form. If it is not possible, say why.

    a. $y = 4^x$
                                   b. $b = 1.5(5)^a$

    c. $2^x = -32$
                                   d. $q = 3(5)^{2k}$

4. Without using your calculator, determine or estimate the value of the variable that makes each equation true.

    a. $\log_2(4) = y$
                                     b. $\log_9\left(\frac{1}{81}\right) = t$

       2
                                            -2

    c. $\log_3(-2) = k$
                                     d. $\log_5(10) = s$

   impossible!

   log w/ negative number don't work..
                                       s between 1 and 2

5. Logarithmic functions are the inverses of basic exponential functions of the form $y = b^x$. That is, both functions show the same relationship between two quantities, but the input and output quantities are switched.

    *a. The graph of $f(x) = \log_4(x)$ is given. Plot the points $(x, f(x))$ when $x = \frac{1}{2}$, 1, 4, and 16.

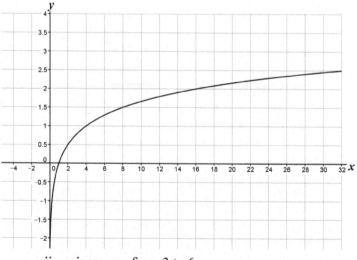

    b. Determine the average rate of change of y with respect to x as:
        i. x increases from 2 to 5
                                   ii. x increases from 3 to 6

---

c. On the axes in part (a), draw the graphs of $g(x) = \log_2(x)$ and $h(x) = \log_{0.5}(x)$. *Create tables of values to help you if necessary.*

Recall that in Module 3 Investigation 7, we defined the irrational number $e = 2.718281828459...$ A logarithm with base e is commonly called the ***natural logarithm*** and is abbreviated "ln."

The Natural Logarithm

The base e logarithm is called ***the natural logarithm*** and is abbreviated ln. For example, $\log_e(150)$ can be written as $\ln(150)$.

Note: Most calculators have two logarithm buttons, log and ln. Even though we can use any positive base (other than one) in a logarithmic function, some calculators only evaluate logarithms for these two bases: base 10 and base e. The "log x" button evaluates $\log_{10}(x)$ *and is referred to as the **common log** of x. The ln x button evaluates* $\log_e(x)$ *and is referred to as the **natural log** of x.*

*6. Convert each of the following to exponential form and evaluate or estimate the value of the unknown. Check your answers using your calculator.

a. $\log\left(\frac{1}{100}\right) = x$

$10^x = \frac{1}{100}$

$x = -2$

b. $\ln(e^2) = k$

$e^k = e^2$

$k = 2$

c. $\ln(7) = t$

$e^t = 7$

$t = 1.945910149$

d. $\log(1000) = s$

$10^s = 1000$

$s = 3$

Recall the Properties of Exponents:

- **Property of Exponents #1:** $b^x \cdot b^y = b^{x+y}$

- **Property of Exponents #2:** $\dfrac{b^x}{b^y} = b^{x-y}$ for $b \neq 0$.

- **Property of Exponents #3:** $\left(b^x\right)^y = b^{xy}$

7. Use your understanding of exponents and the meaning of b^x to justify one of the properties stated above.

Because the result from evaluating a logarithmic expression (and the output of a logarithmic function) represents an exponent, the properties of exponents also apply to logarithms, with the rules of exponents expressed using logarithms and logarithmic form.

Logarithm Properties

Property #1: $\log_b(m \cdot n) = \log_b(m) + \log_b(n)$ for any $b > 0$, $b \neq 1$, and $m, n > 0$

Property #2: $\log_b\left(\dfrac{m}{n}\right) = \log_b(m) - \log_b(n)$ for any $b > 0$, $b \neq 1$, and $m, n > 0$

Property #3: $\log_b(m^c) = c \cdot \log_b(m)$ for any $b > 0$, $b \neq 1$, and $m, n > 0$

The first property of logarithms states $\log_b(m \cdot n) = \log_b(m) + \log_b(n)$. For example, $\log_2(4) + \log_2(8)$ represents the exponent to which 2 is raised to get a result of 4 plus the exponent to which 2 is raised to get a result of 8 (or $2 + 3 = 5$). We could have instead considered $\log_2(4 \cdot 8) = \log_2(32)$ and determined directly that 5 is the exponent to which 2 must be raised to get a result of 32.

8. Use your understanding of logarithmic functions (your knowledge of what the input and output quantities represent) to justify at least one of the logarithmic properties above.

9. Use the properties of logarithms to rewrite each expression. Simplify the result if possible.
    *a. $\ln(x) + \ln(x)$

    $\ln(x^2)$
    OR
    $2 \cdot \ln(x)$

    b. $\log_3(5) + \log_3(2)$

    $\log_3(10)$

    *c. $\log_4(16) - \log_4(4)$

    $\log_4(4) = 1$

    d. $\log_3(2) + \log_5(3)$

    Can't
    Simplify

    *e. $\log_3(9) + \log_3(4) - \log_3(6)$

    $\log_3(6)$

    f. $\log_7(49) - \log_7(-3)$

    not possible
    $\log_7(-16\frac{1}{3})$

*10. Solve each of the following for x.

a. $\log_5(30x^2) = 3$
b. $\log(1.5x) = 0$
c. $\log_3 x + \log_3(2x) = 3$

11. Let $f(x) = 10^x$ and $g(x) = \log(x)$.

a. Complete the tables of values.

x	$f(x)$
-2	
-1	
0	
1	
2	

x	$g(x)$
0.01	
0.1	
1	
10	
100	

b. Evaluate each of the following.

i. $g(f(-2))$
ii. $g(f(0))$
iii. $f(g(1))$

iv. $f(g(10))$
v. $f(g(x))$
vi. $g(f(x))$

c. What do you notice about the relationship between the functions f and g?

*1. Solve each of the following equations for x. Find the exact answer and then use your calculator to approximate the answer to the nearest thousandth (3 decimal places).

a. $4 = 3^x$

$\log_3(4) = 1.261859$

≈ 1.262

b. $22.4 = 17.5(3.4)^x$

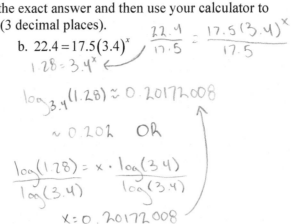

$\frac{22.4}{17.5} = \frac{17.5(3.4)^x}{17.5}$

$1.28 = 3.4^x$

$\log_{3.4}(1.28) \approx 0.20172008$

≈ 0.202 OR

$\frac{\log(1.28)}{\log(3.4)} = x \cdot \frac{\log(3.4)}{\log(3.4)}$

$x = 0.20172008$

2. a. Why is it mathematically valid to "take the log" of both sides of an equation? For example, given $5^x = 800$, why can we write $\log(5^x) = \log(800)$? (*Say more than, "Because we can do anything to one side of the equation as long as we do it to the other side." Think about what the new expressions represent and why they must have the same value.*)

$\log(5^x) = x\log(5)$

$\frac{x \cdot \log(5)}{\log(5)} = \frac{\log(800)}{\log(5)}$

$x = 4.15338279$

$5^x = 800$

$\log_5(800) = 4.15338279$

b. Are there any restrictions on *when* we can "take the log" of both sides of an equation? (Are there equations for which this process is invalid?) Explain.

*3. In 2023, the enrollment at Mainland High School was 1,650 students. Administrators predict that the enrollment will increase by 2.2% each year.

a. Determine when the enrollment at Mainland High School is expected to reach 2,600 students. *You will first need to create a function to model Mainland High School's student population.*

b. If Mainland High School requires 5 teachers per 65 students, how many teachers will be required once the student enrollment reaches 2,600 students?

c. Define a function g that models the number of years since 2023 in terms of the number of students enrolled at Mainland High School.

$$f^{-1}(p) = x$$

$$f^{-1}(p) = \frac{\log(p/218)}{\log(0.981)}$$

inverse of a ↑

*4. *Recall the alpaca problem from Module 3 Investigation 4.* A farm in Canada had 218 alpacas on January 1, 2016. After 8 years, the alpaca population decreased to 187. Assume the number of alpacas decays exponentially.

a. Define a function f that expresses the number of alpacas on the farm in terms of the number of years since January 1, 2016. x = number of years since Jan 1, 2016

$$f(x) = p, \quad p = \text{total alpaca population}$$

$$f(x) = 218(0.981009333)^x$$

b. Using the function defined in part (a), estimate the number of alpacas on the farm on January 1, 2025.

$$218(0.981009333)^9 = 183.4487453$$

about 183 alpacas

c. Use algebraic methods to determine the year when the number of alpacas on the farm reaches 165. Check your answer using a graph or by using the function you defined in part (a).

$$\frac{165}{218} = \frac{218(0.981009333)^x}{218}$$

$$\frac{\log(0.7568807339)}{\log(0.981009333)} = \frac{x \cdot \log(0.981009333)}{\log(0.981009333)}$$

$$14.52798476 = x$$

about 14.5 years → halfway through 2030

*5. At the beginning of 2020, the population of Diamond Bar was 45,000 and the population of Chino Hills was 37,000. At the beginning of 2022, the population of Diamond Bar was 51,750 and the population of Chino Hills was 44,400. Assuming both cities' populations grow at exponential rates, how many years will it take for the population of Chino Hills to surpass the population of Diamond Bar?

6. In 1935, Charles Richter, a seismologist at the California Institute of Technology, invented a method for comparing the magnitudes of earthquakes. Since the amplitudes of seismic waves of big earthquakes can be over a million times as large as the amplitudes of seismic waves of small earthquakes, Richter developed a special scale (now called the Richter Scale) for measuring these earthquake waves. This formula or function takes as input the amplitude of the seismic waves of an earthquake and outputs what he called the earthquake's *magnitude* (the number you will hear reported to describe an earthquake's strength).

The magnitude M of an earthquake with seismic waves of amplitude A is defined to be

$$M = \log_{10}\left(\frac{A}{A_0}\right),$$

where A_0 represents the amplitude of seismic waves for a "standard" earthquake (an amplitude of 1 micron, or 10^{-4} cm).

a. i. Describe the meaning of the ratio $\dfrac{A}{A_0}$ in this context.

ii. Describe the effect of inputting $\dfrac{A}{A_0}$ into the $\log_{10}$ function and what the value of M represents.

b. Let M_1 and M_2 represent the magnitudes (the Richter scale measures) of two earthquakes with seismic waves of amplitudes A_1 and A_2, respectively. Use the formula given above and the properties of logarithms to define a simplified formula for the difference $M_2 - M_1$ in terms of A_1 and A_2.

c. On May 22, 1960, a magnitude 9.5 earthquake struck near Valdivia, Chile. On November 1, 2015, a magnitude 4.1 earthquake struck Phoenix, Arizona. How many times as large were the seismic waves of the earthquake in Chile compared to seismic waves of the earthquake in Arizona?

d. Write a formula that determines the amplitude of the seismic waves of an earthquake in terms of the magnitude of the earthquake.

I. PERCENTAGES AND PERCENT CHANGE (TEXT: S1)

1. You worked a total of 450 minutes at your job yesterday.
   a. What amount of time corresponds to 78% of the total time you spent working yesterday?
   b. If you had worked for 78% of the 450 minutes, what percent of the total time do you still have left to work? What is this amount of time in minutes?

2. You fill a glass with 860 milliliters of water.
   a. What volume of water corresponds to 44% of 860 mL?
   b. If you drank 44% of the water you poured into the glass, what percent of the starting 860 mL remains? How much water is this in mL?

3. A local bicycle shop placed older model bikes on sale last weekend. A customer bought a bike on sale for $299 that had a normal retail price of $495.
   a. How many times as large is the sale price compared with the retail price?

   sale price: $299 _____

   b. *Fill in the blank*: "The sale price is ____% of the retail price."

   retail price: $495 _____

   c. How many times as large is the retail price compared with the sale price?
   d. *Fill in the blank*: "The retail price is ____% of the sale price."

4. Your doctor recommends that you increase or decrease your daily intake of certain vitamins and minerals. For each recommendation described, do the following.
   > i) State the percent change in daily intake.
   > ii) State the number by which we can multiply the original daily intake to find the new recommended daily intake.
   > iii) Find the new recommended daily intake amount.
   a. vitamin D intake: 3μg; increase: 60%
   b. potassium intake: 1500mg; increase: 135%
   c. zinc intake: 30mg; decrease: 45%
   d. calcium intake: 820mg; increase: 25%

5. A store is adjusting the prices of several items. For each price change, do the following.
   > i) State the percent change.
   > ii) State the number by which we can multiply the old price to find the new price.
   > iii) Find the new price.
   a. old price: $29; increase: 5%
   b. old price: $84; increase: 140%
   c. old price: $89.99; decrease: 34%
   d. old price: $6.49; decrease: 22%

6. A record label has seen its annual operating profit decrease in recent years, likely because of greater accessibility of music online. In 2019, the label's operating profit was $135 million. By 2024, the label's operating profit had decreased by 43% (a percent change of −43%).
   a. What was the record label company's operating profit in 2024?
   b. The record label wants to increase its operating profit to $100 million by 2030. By what percent must the label's operating profit increase from its 2024 value to reach $100 million within the next six years?

7. During a recession, Tina had to take a 10% pay cut. Her original salary was $58,240.
   a. What was her salary after the pay cut?
   b. Once the recession was over, Tina's company wanted to increase her pay to her original salary. What percent change in her salary was required to return it to its original $58,240?

II. COMPARING LINEAR AND EXPONENTIAL BEHAVIOR (TEXT: S2)

8. Joni, a biomedical engineer, was offered two positions: one with Company A, and the other with Company B. Company A offered her a starting salary of $156,000 with a 5% guaranteed raise at the end of each of the first five years. Company B offered her $177,000 as her starting salary with a guaranteed raise of $3,500 every year for the first 5 years. She likes both companies and believes she will continue with the company she selects for at least 5 years.
   a. What is the ratio of Joni's salary in one year compared to her salary in the previous year for Company A? Describe how to interpret this ratio.
   b. Fill in the blank: If Joni works for Company A, her salary for any year is ____% of her salary for the previous year.
   c. Define a function f that expresses her salary at Company A in terms of the number of years since she accepts the position, n.
   d. Define a function g that expresses her salary at Company B in terms of the number of years since she accepts the position, n.
   e. After how many years will Joni's salary at Company A overtake her salary at Company B?
   f. Assume she will work for whichever company she chooses for exactly 5 years. A classmate says she should choose Company A because by the time she leaves the company she will have a higher salary. Do you agree? Defend your reasoning.
   g. Find and interpret the following using the functions created in parts (b) and (c).
      i. $f(8)$ ii. $g^{-1}(82,000)$ iii. $g(16)$ iv. $f(20) - f(11)$

9. You are researching jobs in advertising.
   a. You are told that the salary for Job A increases exponentially. The salary for Year 1 is $49,000 while the salary for Year 2 will be $53,066.
      i. How many times as large is the salary in Year 2 compared to the salary in Year 1?
      ii. Fill in the blank: The salary in Year 2 is _____% of the salary in Year 1.
      iii. What is the percent change in salary from Year 1 to Year 2?
      iv. If the salary continues to increase by the same percent each year, what will the salary be in Year 5?
   b. You are told that the salary for Job B also increases exponentially. The salary for Year 1 is $47,500 while the salary for Year 2 will be $52,600.
      i. How many times as large is the salary in Year 2 compared to the salary in Year 1?
      ii. Fill in the blank: The salary in Year 2 is _____% of the salary in Year 1.
      iii. What is the percent change in salary from Year 1 to Year 2?
      iv. If the salary continues to increase by the same percent each year, what will the salary be in Year 5?

10. A chemist monitored the number of living bacteria in a Petri dish after applying a chemical to kill the bacteria. This chemical causes the number of living bacteria of this type to decrease by 12% each hour that elapses after applying the chemical. There were 203 bacteria living when the chemical was applied.
   a. Fill in the blank: At any given time, the number of living bacteria is _____% of the number of living bacteria one hour earlier.
   b. What would the number of living bacteria be after 1 hour? After 2 hours? After 8 hours?
   c. Define a function that expresses the number of living bacteria remaining in the Petri dish, B, as a function of the number of hours since applying the chemical, t.
   d. After how many hours since applying the chemical does the model predict exactly one living bacterium in the Petri dish?
   e. A different type of chemical was applied to another Petri dish of bacteria that caused the number of living bacteria to decrease by 26% each hour. If that Petri dish began with 230 bacteria, define a function that gives the number of living bacteria remaining in this Petri dish, C, as a function of the number of hours that have elapsed since applying the chemical, t.

11. Last year, Jenny invested her birthday money in 3 different penny stocks (penny stocks are stocks that are priced very low). The following functions represent the daily values (rounded to the nearest cent) of each stock over the first 7 days after making the investment. For each investment, do the following.
    i) State the amount of the initial investment.
    ii) Describe how the value of the investment grew over the first 7 days since making the investment using percent change.
    iii) Determine the value of the investment at the end of the 7ᵗʰ day. (*Round your answers to the nearest penny.*)
    a. $h(n) = 25(1.45)^n$ b. $f(n) = 120(3)^n$ c. $g(n) = 275(0.9)^n$

12. There were 24 rabbits living on a 10-acre wildlife preserve on January 1, 2015.
    a. Assuming the number of rabbits doubled each year, determine a function that gives the number of rabbits in the preserve, R, in terms of the number of years elapsed since January 1, 2015, t.
    b. Using the function created in part (a), approximate the number of rabbits in the preserve on January 1, 2022.
    c. What is the percent change per year of the rabbit population?

13. When a teacher asked her algebra class to provide an example of an exponential function, over half of the students offered the function $f(x) = x^2$ because "the function is growing faster and faster."
    a. What is your assessment of this answer?
    b. Compare the growth patterns of $f(x) = x^2$ and $g(x) = 2^x$. First, construct a table of values for each function and create a graph of f and g on the same axes. Then, use the graphs and tables of values to compare their growth patterns.
    c. Compare the growth patterns of $h(x) = x^3$ and $j(x) = 3^x$. First, construct a table of values for each function and create a graph of h and j on the same axes. Then, use the graphs and tables of values to compare their growth patterns.

14. Define a function that models each town's population growth/decline in terms of the number of years since the town was established. Define any variables that you use.
    a. Smallsville starts with 500 people and grows by 10 people per year.
    b. Growsville starts with 500 people and grows by 10% each year.
    c. Shrinktown starts with 500 people and declines by 10% each year.
    d. Littletown starts with 500 people and declines by 10 people per year.

15. Each given function defines the population for a city in terms of the time (in years) since the city was established, t. Write a sentence that describes the city's initial population and growth pattern.
    a. $f(t) = 2000(1.24)^t$ b. $g(t) = 1500 + 20t$ c. $h(t) = 4000(0.68)^t$
    d. $k(t) = 2500 - 40t$ e. $f(t) = 1500(1.4)^{t/2}$

16. A company purchases a new car for their employees to use for $35,000. For accounting purposes, they depreciate the value of the car by 14.5% each year (that is, they declare its value to be 14.5% less for each year that passes).
    a. Using this method of depreciation, what is the value of the car after 2 years?
    b. What is the ratio of the car's value in one year compared to its value in the previous year? Explain the meaning of the value of this ratio.
    c. Define a function that models the value of the car as a function of the number of years since the company purchased it.
    d. When will the value of the car be less than $1000?

17. The U.S. population was about 273.6 million in 1996. Since that time, the population has increased by approximately 1.1% each year.
    a. Define a function *f* that expresses the population of the U.S. in millions, *P*, as a function of the number of years since 1996, *t*.
    b. What was the approximate population of the US in 2010 according to your model?
    c. Assuming the population continues to grow at 1.1% per year, in what year will the US population reach 400 million people?

III. 1-Unit Growth and Decay Factors, Percent Change, and Initial Values (Text: S2)

18. Determine (i) the growth or decay factor, (ii) the percent change, and (iii) the initial value for each of the following exponential functions.
    a. $f(x) = 2^x$
    b. $f(x) = (0.98)^x$
    c. $f(x) = 0.56 \cdot (0.25)^x$
    d. $f(x) = 3 \cdot (1.6)^x$

19. The given tables represent patterns of exponential growth. For each relationship, determine (i) the initial value, (ii) the 1-unit growth/decay factor, (iii) the 1-unit percent change, and (iv) a formula to model the data.
    a.

x	0	1	2	3
$f(x)$	512	384	288	216

    b.

x	1	2	3
$g(x)$	11.2	15.68	21.952

20. For each graph of an exponential function, determine (i) the growth or decay factor, (ii) the percent change, (iii) the initial value, and (iv) the formula.
    a.   b.

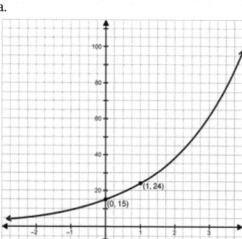
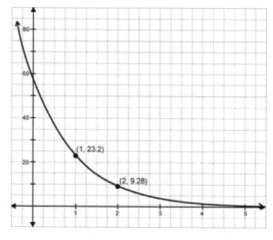

21. An investment of $9000 decreased in value by 2.4% each month.
    a. Determine (i) the initial value of the investment, (ii) the 1-month decay factor, and (iii) the 1-month percent change of the investment.
    b. Define a function *f* that determines the investment's value (in dollars) in terms of the number of months since the initial investment was made.
    c. Determine the value of the investment 12 months after the money was invested.

22. The population of Canada in 1990 was about 27,512,000 and in 2000 it was about 30,689,000. Assume that Canada's population increased exponentially over this time.
    a. Determine (i) the initial value, (ii) the 10-year growth factor, and (iii) the 10-year percent change of Canada's population.
    b. Define a function *f* that defines the population of Canada in terms of the number of *decades* (10-year periods) since 1990.

c. Assuming Canada's growth rate remained consistent, predict the approximate population of Canada in 2040.

23. One year after an investment was made, the amount of money in the account was $1844.50. Two years after the investment was made, the amount of money in the account was $2001.28.
    a. Determine (i) the initial value of the investment, (ii) the investment's 1-month growth factor, and (iii) the investment's 1-month percent change.
    b. Define a function *f* that represents the investment's value (in dollars) in terms of the number of months since the investment was made.
    c. Determine when the value of the investment will reach $4000.

24. The mass of bacteria in a Petri dish was initially measured to be 14 micrograms and increased by 12% each hour over a period of 15 hours.
    a. Determine the 1-hour growth factor of the bacteria.
    b. By what total percent did the bacteria increase during the 15-hour period?
    c. Determine how long it will take for the mass of bacteria to double.

IV. PARTIAL AND *N*-UNIT GROWTH FACTORS (TEXT: S3)

25. A certain strain of bacteria that is growing on your kitchen counter doubles in number every 5 minutes. Assume that you start with only one bacterium.
    a. Define a function that represents the number of bacteria present in terms of the number of 5-minute intervals since the bacteria started growing, *n*.
    b. How many bacteria are present 30 minutes after the bacteria started growing?
    c. Define a function that represents the number of bacteria present *t* minutes since the bacteria started growing.
    d. Define a function that represents the number of bacteria present *h* hours since the bacteria started growing.

26. There were 27 deer in a park reserve on January 1, 2022. Six months later, there were 38 deer. Assume the deer population grows exponentially.
    a. Determine the 6-month growth factor and 6-month percent change for the deer population.
    b. Determine the 1-month growth factor and 1-month percent change for the deer population.
    c. Define a function that represents the number of deer in the park reserve in terms of the number of months since January 1, 2022.
    d. Define a function that represents the number of deer in the park reserve in terms of the number of *6-month intervals* since January 1, 2022.
    e. Use one of your functions to determine when the number of deer in the park reserve reached 125.

27. The population of Egypt at the end of 2002 was 73,312,600, and at the end of 2006 it was 78,887,000. Assume the population of Egypt grew exponentially over this period.
    a. Determine the 4-year growth factor and the 4-year percent change.
    b. Determine the 1-year growth factor and the 1-year percent change.
    c. Define a function that represents Egypt's population in terms of the number of years that have elapsed since the end of 2002.
    d. Rewrite your function from part (c) so that it gives the population of Egypt in terms of the number of *decades* that have elapsed since the end of 2002.
    e. Assuming Egypt's population continued to grow according to this model, how long will it take for the population of Egypt to double?

28. An animal reserve in Arizona had 93 wild coyotes on January 1, 2020. Due to drought, there were only 61 coyotes 3 months later. Assume that the number of coyotes in the reserve decreases (or decays) exponentially.
    a. Find the 3-month decay factor and percent change.
    b. Find the 1-month decay factor and percent change.
    c. Define a function that represents the number of coyotes in terms of the number of months since January 1, 2020. Define any variables you use.
    d. Rewrite your function from part (c) so that it gives the number of coyotes in terms of the number of *years* that have elapsed since January 1, 2020. Define any variables you use.
    e. How long did it take for the number of coyotes to be one-half of the original number?

29. The population of a town increased or decreased by the following percentages. For each situation, find the population's *annual percent change* if we assume each population changed exponentially.
    a. increases by 60% every 12 years
    b. decreases by 35% every 7 years
    c. doubles in size every 6 years
    d. increases by 4.2% every 2 months

30. The number of asthma sufferers in the world was about 84 million in 1990 and 130 million in 2001. Let N represent the number of asthma sufferers (in millions) worldwide t years after 1990.
    a. Define a function that expresses N as a linear function of t. Describe the meanings of the slope and the vertical intercept in the context of the problem.
    b. Define a function that expresses N as an exponential function of t. Describe the meanings of the 1-year growth factor and the vertical intercept in the context of the problem.
    c. The world's population grew by an annual percent change of 3.7% over those 11 years. Did the percent of the world's population who suffer from asthma increase or decrease?
    d. What is the long-term implication of choosing a linear vs. exponential model to make future predictions about the number of asthma sufferers in the world?

31. In the second half of the 20th century, the population of Phoenix, Arizona increased by a large amount. Between 1960 and 2000, the population of Phoenix increased by an average of 2.76% each year. At the end of 2000, the population of Phoenix was 1.32 million people.
    a. Define an exponential function f that represents the population of Phoenix (in millions of people), P, in terms of the number of years after the end of 2000, t.
    b. Sketch a graph of function f.
    c. What input to your function will give the population of Phoenix at the end of 1972? The end of 1983? The end of 1994?
    d. According to your model, approximately how many people lived in Phoenix at the end of 1972? The end of 1983? The end of 1994?
    e. According to your model, in what year did the population reach 1,000,000 people?
    f. On your graph, demonstrate how, for equal increases in t, the value of P changes by larger and larger amounts
    g. Now, define an exponential function g modeling the population of Phoenix (in millions) n years after the end of 1960. Sketch a graph of this function.
    h. How does the function you created in part (g) compare to the original function you created in part (a)? Explain your reasoning.

V. *n*-UNIT GROWTH AND DECAY FACTORS ($n \neq 1$) (TEXT: S3)

32. For each function, find (i) the function's initial value, (ii) the 1-unit growth/decay factor, (iii) the 1-unit percent change, (iv) the 4-unit growth/decay factor, and (v) the 1/3-unit growth/decay factor.
    a. $f(x) = 2^{x/3}$
    b. $f(x) = 5 \cdot (0.98)^{x/4}$
    c. $f(x) = (0.25)^{2x}$
    d. $f(x) = 3 \cdot (1.6)^{5x}$

33. For each exponential relationship, determine (i) the 1-unit growth/decay factor, (ii) the 1-unit percent change, (iii) the 5-unit growth/decay factor, (iv) the ¼-unit growth/decay factor, (v) the function's initial value, and (vi) the function's formula.

a.

x	1	3	5	7
$f(x)$	512	384	288	216

b.

x	2	5	8	11
$g(x)$	8	11.2	15.68	21.952

34. For each exponential relationship, determine (i) the 2-unit growth/decay factor, (ii) the 1-unit growth/decay factor, (iii) the 10-unit growth/decay factor, (iv) the ½-unit growth/decay factor, (v) the function's initial value, and (vi) the function's formula.

a.

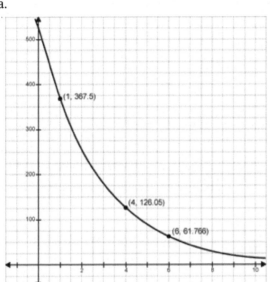

b.

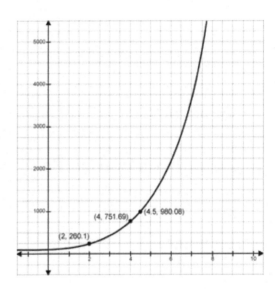

35. Determine the specified growth factors for the following exponential relationship. *The mass of bacteria (in µg) in an experiment t days after its start is given by* $f(t) = 95(4)^{t/5}$.

   a. 5-day growth factor: b. 1-day growth factor:

   c. Select the statement(s) below that describe the behavior of the function above and give reasoning for why the statement(s) you chose are true. *More than one statement might be true.*

      i. An initial mass of 95 *µg* of bacteria quadruples every 1/5-day.

      ii. An initial mass of 95 *µg* of bacteria quadruples every 5 days.

      iii. An initial mass of 95 *µg* of bacteria increases by 50% every day.

36. Determine the specified decay factors for the following exponential relationship. *The number of buffalo in a wildlife preserve t months after its initial measure is given by* $f(t) = 49(0.97)^{2t}$.

   a. 1-month decay factor: b. ½-month decay factor:

   c. Select the statement(s) below that describe the behavior of the function above and give reasoning for why the statement(s) are true. *More than one statement might be true.*

      i. An initial number of 49 buffalo decreases by 3% every ½-day.

      ii. An initial number of 49 buffalo decreases by 3% every 2 days.

      iii. An initial number of 49 buffalo decreases by 5.91% every 1 month.

37. Determine the specified growth/decay factors or percent changes for each exponential relationship.

   a. The population of Jackson is 57,421 and is growing by 4.78% every 4 years.

     i) 8-year growth factor: ii) 8-year percent change in the population:

     ii) 1-year growth factor: iv) 1-year percent change in the population:

b. The amount of a medicine (in milligrams) in your body decreases by 18% every 3 hours.
   i) 24-hour decay factor: ii) 24-hour percent change in the medicine amount:
   iii) 1-hour decay factor: iv) 1-hour percent change in the medicine amount:

38. For each relationship, do the following.
   a. Determine whether the relationship could be linear or exponential.
   b. Fill in the blanks of the tables based on your answer to (a).
   c. Define a function formula that models the data in each table.

Table i			Table ii			Table iii	
Input	Output		Input	Output		Input	Output
–2.0	–10.0		–2.0	0.1111		0.0	132.0
–1.5			–2.0			0.5	
–1.0	–7.5		–2.0	0.3333		1.0	29.04
–0.5			–2.0			1.5	
0	–5.0		–2.0	1.0		2.0	6.3888
0.5			–2.0			2.5	
1.0	–2.5		2.0	3.0		3.0	1.4055
1.5			3.0			3.5	
2.0	0		4.0	9.0		4.0	0.309
						4.5	

VI. COMPOUNDING PERIODS & COMPOUND INTEREST FORMULA (TEXT: S4)

39. The given table illustrates the value of an investment from the end of the 7th compounding period to the end of the 10th compounding period.

Number of compounding periods, p	Investment value (dollars)
7	$7,105.57
8	$7,209.31
9	$7,314.57
10	$7,421.36

   a. Verify that the data in the table represent exponential growth.
   b. How does the investment value change from the end of the 7th to the end of the 9th compounding period? From the end of the 8th to the end of the 10th compounding period?
   c. What is the value of the investment after 3 compounding periods? After 20 compounding periods?
   d. Define a function f that expresses the investment's value as a function of the number of compounding periods, p. *Note: p is defined to be the whole number values* $\{0,1,2,3,...\}$. Explain the meaning of each of the values in your function.
   e. Determine the number of compounding periods needed for the investment to reach $400,000.

40. For each of the following accounts, determine the percent change per compounding period. Give your answer in both decimal and percentage form.
   a. interest rate of 5% compounded monthly   b. interest rate of 6.7% compounded quarterly
   c. interest rate of 3.2% compounded daily   d. interest rate of 8% compounded each hour

41. For each of the accounts in Exercise #40, determine (i) the growth factor per compounding period, (ii) the annual growth factor of the account value, and (iii) the account value's annual percent change (APY).

42. If you invest $1,200 in a CD with an interest rate of 3.5% compounded monthly, the following expression will calculate the value of the CD in 6 years: $1200\left(1+\frac{0.035}{12}\right)^{12(6)}$.

   Explain what each part of the expression represents in this calculation.
   a. $\frac{0.035}{12}$ b. $1+\frac{0.035}{12}$ c. $12(6)$ d. $\left(1+\frac{0.035}{12}\right)^{12}$ e. $\left(1+\frac{0.035}{12}\right)^{12(6)}$

43. Write an expression that would calculate the value of each of the following accounts after 15 years.
   a. $8100 is invested at an interest rate of 4.8% compounded semiannually (twice per year).
   b. $16,000 is invested at an interest rate of 3.5% compounded daily.

44. An investment of $10,000 with Barnes Bank earns a 2.42% interest rate compounded *monthly*.
    a. Define a function that gives the investment's value (in dollars) as a function of the number of years since it was opened.
    b. Determine the investment's value after 20 years.
    c. Determine the annual growth factor and annual percent change (APY).
    d. Determine how long it will take for the investment value to double.
    e. Another bank says they will pay you the same interest rate (2.42%) but they will compound the interest daily. What would be the value of a $10,000 investment with this bank if you left it there for 20 years? Compare this answer to your account with Barnes Bank.

45. You decide to deposit $1000 in an investment account and leave it there for 10 years.
    a. Which of the three different accounts would you choose to invest your $1000? Provide calculations for each account and justify your reasoning.
        • Account #1 pays 14% interest each year, compounded annually (once per year).
        • Account #2 pays 13.5% interest per year, compounded monthly.
        • Account #3 pays 13% interest per year, compounded weekly.
    b. Describe how your money increases as time increases for each account. Be sure to incorporate the annual growth factor and the annual percent change into your description.

46. Karen received an inheritance from her grandparents and wants to invest the money. She is offered the following two options of accounts to invest into:
        • 4.5% interest rate, compounded semi-annually
        • 4.3% interest rate, compounded daily
    a. Which option should she choose? Explain your reasoning using the accounts' APY values.
    b. If Karen decides on the first option, by what *percentage* will her investment have *increased* after 10 years?

47. John decides to start saving money for a new car. He knows he can invest money into an account which will earn an interest rate of 6.5% compounded weekly, and he would like to have $10,000 saved after 5 years.
    a. How much money will he need to invest into the account now so that he has $10,000 after 5 years?
    b. Determine the APY (Annual Percent Yield) for the account.
    c. Determine the 5-year percent change for the account.

VII. INVESTMENT ACTIVITY: FOCUS ON FORMULAS AND MOTIVATING *e* (TEXT: S4)

48. $4500 is initially invested into an account with an interest rate of 7%. Determine the value of the account after each number of years and the APY for each compounding frequency.

Year	Compounded Monthly	Compounded Daily	Compounded Continuously
1			
2			
4			
10			
30			
APY			

49. Determine the value (in dollars) of each of the following accounts after 14 years.
    a. Initial investment is $4000 with an interest rate of 7.1% compounded continuously.
    b. Initial investment is $750 with an interest rate of 3.3% compounded continuously.
    c. Initial investment is $2000 with an interest rate of 5% compounded continuously.

50. $2000 is initially invested in an account with an interest rate of 3.4% compounded continuously.
    a. Determine the account's value at the end of 5 years.
    b. Write a function that models the value of the account at the end of t years.
    c. What is the annual percent change (APY) of the account? What does this value represent?
    d. What is the percent change over 10 years for this account?

51. $18,000 is initially invested in an account with an interest rate of 5.1%.
    a. Does it make a bigger impact going from compounding interest annually to monthly, or going from compounding interest daily to continuously?
    b. What is the annual percent change (APY) for each of the four compounding methods listed in (a)?
    c. If interest is compounded continuously, how much interest does the account earn over the first 10 years?

52. An initial population of 32 million people increases at a continuous percent rate of 1.9% per year since the year 2000.
    a. Determine the function that gives the population in terms of the number of years since 2000.
    b. Determine the population in the year 2019.
    c. What is the annual growth factor for this context? Explain two ways to determine this value.
    d. What does your answer to part (c) represent in this context?

53. Carbon-14 is used to estimate the age of an organic compound. Over time, carbon-14 decays at a continuous percent rate of 11.4% per thousand years from the moment the organism containing it dies. Carbon-14 is typically measured in micrograms.
    a. Define a function that gives the amount of carbon-14 remaining at any moment in a piece of wood that starts out with 150 micrograms of carbon-14. (*Remember that carbon-14's decay rate is per thousand years.*)
    b. Construct a graph of this function. Be sure to label your axes!
    c. What is the percent change every 1000 years?
    d. What is the percent change every 5000 years? Explain your reasoning.

VIII.   THE INVERSE OF AN EXPONENTIAL FUNCTION (TEXT: S5)

Log Property #1: $\log_b(m \cdot n) = \log_b(m) + \log_b(n)$ for any $b > 0$, $b \neq 1$ and $m, n > 0$

Log Property #2: $\log_b\left(\frac{m}{n}\right) = \log_b(m) - \log_b(n)$ for any $b > 0$, $b \neq 1$ and $m, n > 0$

Log Property #3: $\log_b\left(m^c\right) = c \cdot \log_b m$ for any $b > 0$, $b \neq 1$ and $m, n > 0$

54. Estimate the value of each logarithmic expression. Explain why your estimate makes sense.
    a. $\log_4(60)$ b. $\log_3(143)$ c. $\log_8\left(\frac{1}{64}\right)$ d. $\log_9(27)$

55. Find the unknown value in each of the following equations. *Estimate if necessary. For parts (g) and (h), assume z is a positive number.*
    a. $\log_3(20) = y$ b. $2\left(\log_5(625)\right) = y$ c. $\log_5(1) = y$ d. $\log_2(30) = y$
    e. $\log_2(x) = 4.2$ f. $\log_7(x) = -2$ g. $\log_z(343) = 3$ h. $\log_z(1) = 0$

56. For each of the following expressions:
    • Write the expression as a single logarithm.
    • Evaluate the expression as a single number or estimate the expression's value.
    a. $\log_5(6.25) + \log_5(100) + \log_5(25)$ b. $\log_4\left(\frac{1}{32}\right) + \log_4\left(\frac{1}{8}\right)$ c. $2 \cdot \log_6(12) + \log_6(4)$
    d. $\ln(6) + \ln(3) - \ln(2)$ e. $\log_5(6) - \log_5(100)$ f. $\frac{1}{2}\left(\log_5(8) + \log_5(3)\right)$

57. Solve each of the following for x. (*Hint: Use the properties discussed in class or your understanding of logarithms.*)

    a. $\log(0.01x) = 0$ b. $\log_5(25x^2) = 6$ c. $\ln\left(\frac{x}{10}\right) = 4$

58. Solve each of the following for x.

    a. $\log_2(2 + x) + \log_2(7) = 3$ b. $\ln(3x^2) - \ln(5x) = \ln(x + 9)$

59. Rewrite each of the following as a sum and/or difference of logarithms. Expand as much as possible.

    a. $\log_7\left(\frac{4}{y}\right)$ b. $\log_2\left(x^4 \cdot 12\right)$ c. $\log_5\left(\frac{10x^3}{y^5}\right)$

60. Rewrite each of the following exponential equations in logarithmic form.

    a. $y = 11^x$ b. $y = 1.7(3.2)^t$ c. $y = 200(1.0027)^{12t}$

61. Graph each of the following functions.

    a. $f(x) = \log_3(x)$ b. $g(x) = \log_5(x)$

62. a. Sketch the following two functions on the same set of axes: $f(x) = \log(x)$ and $g(x) = \log_5(x)$.

    b. For what values of x is $\log(x) > \log_5(x)$?

IX. Solving Exponential and Logarithmic Equations (Text: S5)

63. The value of an investment (in dollars) is represented by $f(t) = 4186.58(1.025)^t$ where t is the number of years since January 1, 2020. Algebraically, determine when the investment will reach $1,000,000.

64. An initial amount of 120 mg of caffeine is metabolized in Sam's body and the amount in Sam's bloodstream decreases at a continuous percent rate of 21% per hour.

    a. Define an exponential function that gives the amount of caffeine remaining in Sam's bloodstream after t hours.

    b. How many hours will it take for the amount of caffeine in Sam's bloodstream to reach half of the initial amount? (*Solve this both graphically and symbolically to verify your answers.*)

65. An initial investment of $6000 is made in an account with an interest rate of 4.7%.

    a. If interest is compounded monthly, how many years will it take for the account balance to be $10,290.32? Solve algebraically and check your answer.

    b. If interest is compounded continuously, how many years will it take for the account balance to be $9,510.15?

66. The amount of medicine in a patient's bloodstream for reducing high blood pressure decreases at a continuous percent rate of 27% per hour. This medicine is effective until the amount in the bloodstream drops below 1.2 mg. A doctor prescribes a dose of 85 mg.

    a. Define function A that represents the amount of medicine remaining in the patient's bloodstream t hours after the dose was administered.

    b. About how long will it be until 45 mg of medicine remain in the patient's bloodstream? 1.2 mg?

67. The rate at which a wound heals can be modeled by the exponential function $f(n) = Ie^{-0.1316n}$ where I represents the initial size of the wound (in square millimeters) and $f(n)$ represents the size of the wound after n days. This function assumes no infection is present and no antibiotic ointment is used to speed healing.
    a. Suppose you scrape your knee and get a wound 300 square millimeters in size. Define a function to model the size of the wound with respect to time.
    b. How large will the wound be after one week?
    c. How long will it take for the wound to be 20% of its original size?
    d. You want to know how long it will take for the wound to shrink from the size you determined in part (c) to 20% of the size you determined in part (c). How will this amount of time compare to the amount of time it took to reduce the wound to 20% of its original size? Explain your reasoning.

68. Given the function $f(t) = 10(0.71)^t$, complete the following.
    a. What is the 1-unit percent change in the function's value?
    b. Convert $f(t) = 10(0.71)^t$ into the equivalent form $f(t) = 10e^{rt}$.

69. The town of Gilbertville increased from a population of 3,562 people at the end of 1970 to a population of 9,765 at the end of 2000.
    a. Define an exponential function that models the town's population as a function of the number of years since the end of 1970.
    b. What is the annual percent change in the town's population?
    c. Use your function to predict the town's population at the end of 2049 if this growth rate continues.
    d. According to your function, when will the town's population reach 40,000 people? (*Answer this question both graphically and symbolically.*)
    e. After how many years from *any* reference year will the population triple?

This investigation contains review and practice with important skills and procedures you may need in this module and future modules. Your instructor may assign this investigation as an introduction to the module or may ask you to complete select exercises "just in time" to help you when needed. Alternatively, you can complete these exercises on your own to help review important skills.

Factoring Variable Expressions
Use this section prior to the module or with/after Investigation 1.

Factoring is the process of rewriting a number as a product of factors. For example, we can rewrite 10 as $5 \cdot 2$. The value is still the same but we've written the number as a product.

What if the number is the value of a quantity that varies in value? For example, the expression $3x^2 + 12x$ varies in value and, as written, is evaluated as the sum of two variable expressions ($3x^2$ and $12x$). But this expression can be written instead as a product of two variable expressions, $3x(x + 4)$. For ALL values of x, $3x^2 + 12x$ and $3x(x + 4)$ have the same value.

1. Evaluate the expressions $3x^2 + 12x$ and $3x(x + 4)$ for each given value of x to verify their values are identical in each case.
    a. $x = 7$ b. $x = 43$ c. $x = -9$

In Exercises #2-4, do the following.
  (i) Rewrite the expression in factored form.
  (ii) Pick a value for x and use it to evaluate the original expression and your factored form to demonstrate that your answer to part (i) represents the same value as the original expression written as a product of two numbers.

2. $5x + 35$ 3. $6x^2 - 13x$ 4. $3x^2 + 12x$

In Exercises #5-7 you are given a variable expression written in factored form. Use the distributive property to rewrite each expression in expanded form.

5. $6x(x + 7)$

6. $6y(3y - 2x)$

7. $a(ab^2 + a^3c)$

Expanding Binomial Products

Use this section prior to the module or with/after Investigation 3.

When a product involves two variable expressions with two or more terms, such as $(x + 7)(x + 11)$, writing the expanded form can sometimes be tricky. One very useful way to think about the product is to use an area model.

To demonstrate the idea, let's think about the number 21, which could be written as a product $(7)(3)$, and even broken down further and written as a product of two sums like $(3 + 4)(2 + 1)$. Let's visualize this product as representing a rectangular area of 21 square units. *See diagram to the right.*

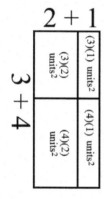

If we break down this area into sub-rectangles, we can see that the entire area is made up of four sub-areas measuring $(3)(2)$ square units, $(3)(1)$ square units, $(4)(2)$ square units, and $(4)(1)$ square units. *See diagram to the left.*

Thus, $(3 + 4)(2 + 1)$ is equivalent to $(3)(2) + (3)(1) + (4)(2) + (4)(1)$.

We can use the same basic idea to visualize the product of $(x + 2)(x + 3)$. *See diagram to the right.*

Using this diagram, it should be clear that $(x + 2)(x + 3)$ is equivalent to $x^2 + 3x + 2x + (2)(3)$. This expression can be simplified to $x^2 + 5x + 6$. This is the expanded (and simplified) form.

In Exercises #8-10, rewrite each product in expanded form by first drawing an area diagram.

8. $(x + 6)(x + 2)$

9. $(x + 3)(x + 5)$

10. $(x + 1)(x + 4)$

In Exercises #11-13, rewrite each product in expanded form. *You do not need to draw an area diagram but you can if it helps you. Remember that something like $(x + 3)^2$ is shorthand for $(x + 3)(x + 3)$.*

11. $(x + 8)(x + 3)$ 12. $(2x + 3)(3x + 7)$ 13. $(x + 8)^2$

In Exercises #14-16, rewrite each product in expanded form. *We recommend rewriting subtraction as addition with a negative to avoid sign errors. For example, rewrite $x - 5$ as $x + (-5)$. You do not need to draw an area diagram but you can if it helps you. Remember that something like $(x - 3)^2$ is shorthand for $(x - 3)(x - 3)$.*

14. $(x - 3)(x + 5)$ 15. $(x + 2)(3x - 1)$ 16. $(2x - 3)^2$

17. a. Your friend said that $(a + b)^2$ can be written as $a^2 + b^2$. Is he right? Justify your answer using an area diagram.

   b. Use an area diagram to help you write the general expanded form for any product $(a - b)^2$.

Factoring Polynomial Expressions
Use this section prior to the module or with/after Investigation 3.

If we want to write a polynomial expression like $x^2 + 7x + 10$ in factored form, we can use an area diagram to help us organize our thinking. From our work above, it should be clear that the top left rectangle in our area diagram will have an area of x^2 square units (with side lengths of x and x units) and the bottom right rectangle must have an area of 10 square units.

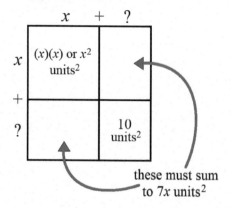

What we need to figure out are the dimensions of the bottom right rectangle that will 1) make its area 10 square units and 2) make the areas of the other two rectangles sum to $7x$ square units.

Our options include any pair of numbers with a product of 10 (which includes $10 \cdot 1$ and $5 \cdot 2$). These options are shown below.

$$(x+10)(x+1) \qquad \text{or} \qquad (x+5)(x+2)$$

18. a. Which of the options is "correct"? What does that tell us about the factored form of $x^2 + 7x + 10$?

   b. Choose a value for x and evaluate the expression $x^2 + 7x + 10$. Then use the same value to evaluate the factored form you chose in part (a) to show that it represents the same value of $x^2 + 7x + 10$ but written as a product of two variable expressions.

   c. Repeat part (b) with a different value for x.

In Exercises #19-21, rewrite each expression in factored form. For example, $x^2 + 9x + 14$ would be rewritten as $(x+7)(x+2)$. *Draw an area diagram if it helps you complete the process.*

19. $x^2 + 8x + 12$ 20. $x^2 + 9x + 8$ 21. $x^2 + 2x - 15$

Solving Polynomial Equations
Use this section prior to the module or with/after Investigation 4.

Consider the polynomial equation $x^2 + 8x + 7 = 0$. The solutions are the values of x such that the expression $x^2 + 8x + 7$ has a value of 0. There are several ways to solve these equations. Let's start by factoring the expression.

$$x^2 + 8x + 7 = 0$$
$$(x + 7)(x + 1) = 0$$

We now have a product of two numbers (the number $x + 7$ and the number $x + 1$) that evaluates to 0. What does that tell us? The only way a product of two numbers can be 0 is if one of the two numbers in the product is 0. In other words, if $ab = 0$, then either $a = 0$ or $b = 0$. This is known as the ***zero-product property***.

What that means in this context is that either $(x + 7) = 0$ or $(x + 1) = 0$.

$$x^2 + 8x + 7 = 0$$
$$(x + 7)(x + 1) = 0$$
$$x + 7 = 0 \quad \text{or} \quad x + 1 = 0$$
$$x = -7 \quad \text{or} \quad x = -1$$

So, either $x = -7$ or $x = -1$. These are the only values that could make $x^2 + 8x + 7$ have a value of 0.

22. Substitute $x = -7$ and $x = -1$ into the expression $x^2 + 8x + 7$ to show that each produces a value of 0.

In Exercises #23-25, solve each equation by factoring and using the zero-product property. Check your answer by graphing. *To do this, graph a function f whose outputs are defined by the polynomial expression and demonstrate that the function value is 0 for each x-value solution.*

23. $x^2 - 13x + 30 = 0$ 24. $x^2 + 7x - 18 = 0$ 25. $2x^2 + 5x + 3 = 0$

Sometimes the equation is not set equal to 0, meaning we can't use the zero-product property. However, we can rewrite the equation so that the zero-product property applies. See the following example.

$$x^2 + 9x + 13 = 4x + 7 \qquad \text{(subtract } 4x \text{ and 7 from both sides)}$$
$$x^2 + 5x + 6 = 0$$
$$(x + 3)(x + 2) = 0$$
$$x + 3 = 0 \quad \text{or} \quad x + 2 = 0$$
$$x = -3 \quad \text{or} \quad x = -2$$

In Exercises #26-27, solve each equation by first rewriting it so that the equation is equal to 0.

26. $x^2 + 5x + 2 = 3x + 10$

27. $x^2 - 12x + 24 = -2x + 3$

If the equation is quadratic (and written in the form $ax^2 + bx + c = 0$), then the *x*-values that make the expression evaluate to 0 (the solutions to the equation) can be represented by $\boxed{x = -\frac{b}{2a} \pm \frac{\sqrt{b^2 - 4ac}}{2a}}$. This is called the **quadratic formula** and represents the solutions to equations of the form $ax^2 + bx + c = 0$.

Note that "$\pm$" indicates that the quadratic formula can produce two solutions.

$$2x^2 - 5x + 1 = 0 \quad (\text{so } a = 2, b = -5, c = 1)$$

$$x = -\frac{b}{2a} \pm \frac{\sqrt{b^2 - 4ac}}{2a}$$

$$x = -\frac{-5}{2(2)} \pm \frac{\sqrt{(-5)^2 - 4(2)(1)}}{2(2)}$$

$$x = -\frac{-5}{4} \pm \frac{\sqrt{25 - 8}}{4}$$

$$x = \frac{5}{4} \pm \frac{\sqrt{17}}{4}$$

The solutions are $x = \frac{5}{4} + \frac{\sqrt{17}}{4}$ and $x = \frac{5}{4} - \frac{\sqrt{17}}{4}$.

In Exercises #28-30, solve each equation using the quadratic formula. *Be very careful and pay attention to the signs of each number and the signs in the formula.*

28. $2x^2 + 6x + 3 = 0$

29. $x^2 + 4x - 1 = 0$

30. $5x^2 - 2x - 1 = 0$

*1. Examine the given bottle and imagine this bottle filling with water. While imagining the bottle filling with water, consider how the height of the water in the bottle changes for equal changes in volume.

    a. Draw a rough sketch of the graph representing the height of water in the bottle in terms of the volume of water in the bottle. *Label your axes. Your graph does not need to be exact—a rough sketch is good enough!*

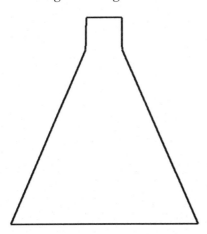
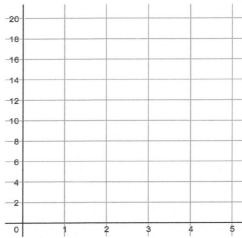

    b. Based on your graph, describe how the height of the water in the bottle changes as the volume of water in the bottle increases by successive equal amounts (e.g., 1 cup) throughout the domain.

*2. a. Now, use the applet provided in the PowerPoint to determine the height of the water in the bottle for the 7 different volume values noted on the applet (and appearing in the table).

Note that it is useful to select values of the volume by repeatedly adding the same amount of water for each section of the bottle when determining the water's height. Explain why this strategy is helpful.

Volume of Water (cups)	Height of Water (cm)	Change in the Water's Height (cm)
0	0	
1	2	
2		
3		
4		
4.2		
4.4		

    b. Use the values in your table to draw a more accurate graph of the height of the water in the bottle in terms of the volume of water in the bottle.

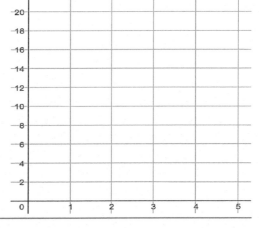

*3. The given graph represents the *height of water in a second bottle* in terms of the *volume of water in the bottle*. Use this graph to complete the following exercises.

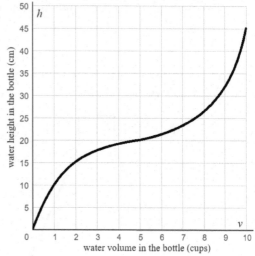

a. Consider how the water's height changes as the volume of water increases from 0 to 2.5 cups, and from 2.5 to 5 cups.
   i. Over which of these two intervals does the water's height increase more? Explain.

   ii. Illustrate (by drawing arrows on the given graph) the height changes that correspond with the volume changing from 0 to 2.5 cups and then from 2.5 to 5 cups.

   iii. What does the information in parts (i) and (ii) suggest about the shape of the bottom end of the bottle?

b. Consider how the water's height changes as the volume increases from 5 to 7.5 cups, and from 7.5 to 10 cups.
   i. Over which of these two intervals does the water's height increase more? Explain.

   ii. Illustrate (by drawing arrows on the given graph) the height changes that correspond with the volume changing from 5 to 7.5 cups and then from 7.5 to 10 cups.

   iii. What does the information in parts (i) and (ii) suggest about the shape of the upper end of the bottle?

c. Note that the volume-height graph changes from curving downward to curving upward when 5 cups have been added to the bottle. What does this information convey about the bottle's shape?

d. Draw a picture of a bottle that would produce the volume-height relationship conveyed in the given graph. Label landmarks on your bottle, and on the graph in part (a), where the function's behavior changes in important ways.

The given graph (from Investigation 1) represents the relationship between the height (in cm), h, and volume of water (in cups), v, in a bottle. Call this function relationship f.

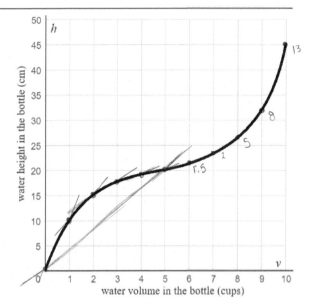

*1. a. On the graph, plot the points $(0, f(0))$ and $(5, f(5))$.

   b. Draw a line that passes through these points.

   c. Determine the slope of the line you drew by estimating values for $f(0)$ and $f(5)$ on the graph.

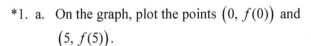

$$\frac{20-0}{5-0} = 4 \text{ cm/cup}$$

   d. What does this slope represent in this context?

The average rate of change between the two points

It's clear that, as the volume of water in the bottle changes from 0 cups to 5 cups, the water height does NOT change at a constant rate with respect to the water volume. However, the slope of the line connecting the two points represents *a constant rate of change* (of the water height with respect to the water volume) that achieves the same net change in the water height that the actual function achieved for the same change in volume. This is the *function's average rate of change over that interval.*

   e. It is often useful to determine the average rate of change of a function over smaller and smaller intervals of the function's domain. Discuss with your classmates why you think this might be useful. Provide your answer below.

   f. Construct 5 line segments on the graph of f by connecting the points:

     i) $(0, f(0))$ and $(1, f(1))$ ii) $(1, f(1))$ and $(2, f(2))$ iii) $(2, f(2))$ and $(3, f(3))$

     iv) $(3, f(3))$ and $(4, f(4))$ v) $(4, f(4))$ and $(5, f(5))$

   g. Discuss the following questions with your classmates, then provide your answer.

     i. How does the *function's average rate of change* (over 1-cup intervals of the function's domain) *change* as the volume of water in the bottle increases from 0 to 5 cups?

It decreases

ii. What does the pattern of change in the function's average rate of change values tell you about the shape of the function's graph? Is the graph increasing or decreasing? Is the graph concave up (cupped upward) or concave down (cupped downward)?

iii. What does the pattern of change in the function's average rate of change values tell you about how the height of the water in the bottle and volume of water in the bottle are changing together?

h. Is the function's graph concave up (cupped upward) or concave down (cupped downward) on the interval $5 < v < 10$? What does this convey about how the water height and the water volume are changing together?

Concave up

As volume increases, height increases more and more

i. At what value(s) of v does the graph change concavity? (Note that a point $(v, f(v))$ where the graph changes concavity is called an **inflection point**.)

$(5, f(5))$

Rates of change are measurements of how a function's output quantity varies with respect to changes in the value of its input quantity. **Concavity** is a measurement of how a function's <u>rate of change</u> itself varies with changes in the value of a function's input quantity.

Concavity

For any function f, imagine looking at an interval of the domain from $x = a$ to $x = b$ and dividing it into any number of <u>equal-sized</u> subintervals.

- f is said to have **positive concavity** (or to be **concave up**) on the interval (a, b) if the function's average rate over successive intervals always increases.
- f is said to have **negative concavity** (or to be **concave down**) on the interval (a, b) if the function's average rate over successive intervals always decreases.

In Exercise #1, the function changed from having negative concavity when the volume was less than 5 cups to having positive concavity when the volume was greater than 5 cups. The point $(5, f(5))$ is thus an ***inflection point*** for the function.

Inflection Point

A continuous function f has an inflection point at $\big(a, f(a)\big)$ if the function has a different concavity when $x < a$ compared to when $x > a$.

---

*2. a. Illustrate a change in x from –6.3 to –3.9 on the number line below.

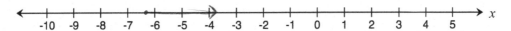

b. True or False: The value of x is decreasing as x changes from –6.3 to –3.9. Justify your response.

False, it's a positive increase

*3. A 10-foot ladder is leaning against a wall in the vertical position. A carpenter then pulls the base of the ladder away from the wall at a constant rate until the top of the ladder hits the floor.

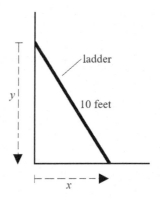

a. After considering the situation and discussing it with your classmates, make a conjecture about the speed at which the **top of the ladder** slides toward the floor. Does it slide at a constant speed, or does it slow down or speed up? Justify your claim and explain how we could test your conjecture.

It speeds up!

b. Construct a function f that represents the distance of the top of the ladder from the floor (in feet), y, in terms of the distance of the bottom of the ladder from the wall (in feet), x. (*Hint: Use the Pythagorean Theorem.*)

c. Use *f* to complete the given table of values.

Distance of the bottom of the ladder from the wall (in feet) x	Distance of the top of the ladder from the floor (in feet) $y = f(x)$	Change in the distance of the top of the ladder from the floor (in feet) $\Delta f(x)$
0	10	
2		
4		
6		
8		
10	0	

d. For each *x*-interval specified in the given table, determine *f*'s average rate of change (the average rate of change of *the ladder's height above the ground* with respect to *the ladder's distance from the wall*). Feet per foot   Feet the ladder falls per foot the bottom is moved

e. Examine the patterns of change in these average rate of change values and explain what this conveys about how the top of the ladder slides toward the floor. Does it slide at a constant speed, or does it slow down or speed up? Explain.

x- interval	Average rate of change of $f(x)$ with respect to x
0 to 2	- 0.101
2 to 4	- 0.3165
4 to 6	-0.5825
6 to 8	-1
8 to 10	-3

It speeds up

ROC is negative bc height is decreasing

Concave down, negative concavity

f. Construct a graph of function *f*. Is the graph concave up or concave down? Justify your answer.

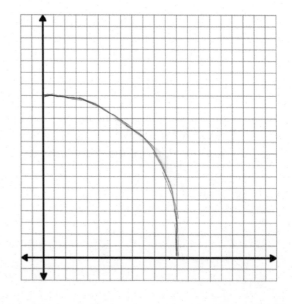

Now that we have introduced and explored the idea of concavity, let's look more closely at how to identify intervals where a function is ***concave up*** or ***concave down*** by looking at its graph and *why* the concavity is ***positive*** or ***negative*** over those intervals. Graphs of functions over intervals with positive concavity "curve up" or "cup upward" as *x* increases. Graphs of functions over intervals with negative concavity "curve down" or "cup downward" as *x* increases. Examples are shown below, along with line segments whose slopes represent the functions' average rates of change over consecutive intervals.

Example I: positive concavity ("concave up") on the entire domain

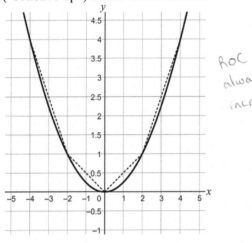

ROC always increasing

Example II: positive concavity ("concave up") on the entire domain

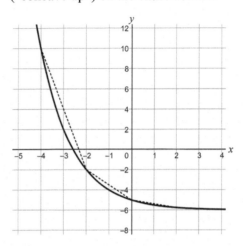

AROC less negative

Example III: negative concavity ("concave down") on the entire domain

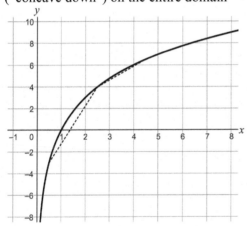

ROC always decreasing

Example IV: negative concavity ("concave down") on the entire domain

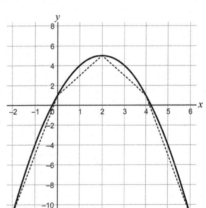

4. *a. For Example I, explain why the function has positive concavity over its entire domain (from what we can see). Then explain how you can "see" the graph cupping upward as *x* increases.

RoC Always getting more positive

b. For Example II, explain why the function has positive concavity over its entire domain (from what we can see). Then explain how you can "see" the graph cupping upward as x increases.

*c. For Example III, explain why the function has negative concavity over its entire domain (from what we can see). Then explain how you can "see" the graph cupping downward as x increases.

d. For Example IV, explain why the function has negative concavity over its entire domain (from what we can see). Then explain how you can "see" the graph cupping downward as x increases.

5. For each of the given functions, do the following.
    i. State any interval(s) of the domain over which it appears the function has positive concavity.
    ii. State any interval(s) of the domain over which it appears the function has negative concavity.
    iii. Estimate any inflection points for the function.

  *a. $f(x) = x^3 - 6x^2 + 9x + 10$

  b. $g(x) = -3 \cdot \sqrt{x+4} + 6$

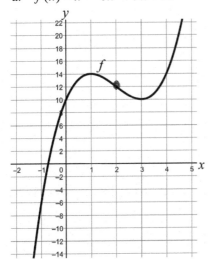
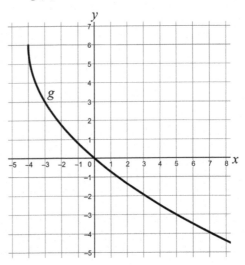

6. Graph each of the following functions using a graphing calculator or graphing software. For each, do the following.
    i. Estimate the interval(s) over which the function's graph is concave up.
    ii. Estimate the interval(s) over which the function's graph is concave down.

  a. $f(x) = x^3$

  b. $g(x) = -2x^2 + 4x - 1$

  c. $h(x) = -x^3 + 6x$

  d. $j(x) = 3^x$

*7. At 5:00 p.m. Karen started walking from the grocery store to her house.

   a. Complete the table of values using the information provided.

Change in the number of minutes since Karen started walking Δt	Number of minutes since Karen started walking t	Karen's distance from her house (in feet) d	Change in Karen's distance from her house (feet) Δd	Average rate of change of Karen's distance from her house with respect to the number of minutes since she started walking
	0	440		
			–45	
	0.5			
			–67	
	1			
			–95	
	1.5			
	2	38		

   b. Sketch a graph for function f, Karen's distance from her house (in feet), d, as a function of the number of minutes since Karen started walking, t, based on the information given. (*Be sure to label your axes.*)

   c. Based on the information given, does f have positive concavity ("concave up"), negative concavity ("concave down"), or some combination of both on the interval $0 < t < 2$? Make sure you can justify your answer.

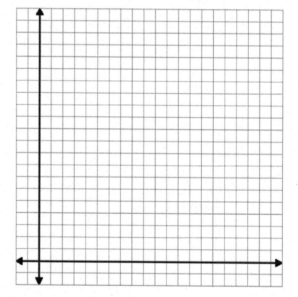

   d. Describe how the quantities *number of minutes since Karen started walking* and *Karen's distance from her house (in feet)* change together.

*1. Regal Theater puts on community theater productions each season. They used several years' worth of data to create the given model representing the number of tickets they expect to sell for Friday evening shows in terms of the length of the production.

play length (hours) x	**Regal Theater** number of tickets they expect to sell to Friday evening shows $g(x)$
1	119
1.5	139
1.75	145
2	149
2.25	150
2.5	149
3	139
3.5	119
4	89

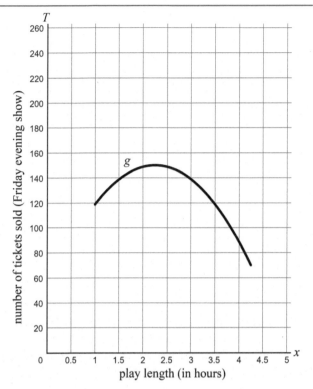

a. What does the ordered pair (3, 139) represent in the context of this situation?

b. What is *g*'s concavity over the interval $1 < x < 4$? What does this tell you about the situation?

*2. Regal Theater contacted a similar theater in another city (Player Productions) to compare models. Player Productions said their model was virtually identical to the model created by Regal Theater. The only difference is that they sold 30 fewer tickets each Friday night compared to Regal Theater.

a. Complete the table of values for Player Productions' model.

b. Describe how the outputs of functions *g* and *h* compare for any input value *x*.

play length (hours) x	**Player Productions** number of tickets they expect to sell to Friday evening shows $h(x)$
1	
1.5	
1.75	
2	
2.25	
2.5	
3	
3.5	
4	

c. Draw a graph of h on the axes given in Exercise #1.

d. How do the average rates of change of h on the intervals $1 \le x \le 2.25$ and $2.5 \le x \le 4$ compare to the average rates of change of g on the same intervals? Why?

e. Use function notation to express the outputs of h in terms of the outputs of g.

*3. A third theater company (Actor's Guildhouse) found a similar trend; however, they sell 1.5 times as many tickets to their Friday shows compared to Regal Theater.

a. Complete the given table of values for Actor's Guildhouse's model.

b. Describe how the outputs of functions g and f compare for any input value x.

play length (hours) x	**Actor's Guildhouse** number of tickets they expect to sell to Friday evening shows $f(x)$
1	
1.5	
1.75	
2	
2.25	
2.5	
3	
3.5	
4	

c. Draw a graph of f on the axes given in Exercise #1.

d. Use function notation to express the outputs of f in terms of the outputs of g.

*4. The number of tickets to a Friday performance in terms of the play length (in hours) for a fourth theater company (Stage Left) is modeled by function j.

a. If $j(x) = g(x - 0.5)$, how do the expected ticket sales for a Friday performance compare at Stage Left and Regal Theater? [*Hint: Which function needs a larger input value to produce the same output value?*] How would the graphs of the functions compare?

b. If instead $j(x) = 2g(x) - 15$, how do the expected ticket sales for a Friday performance compare at Stage Left and Regal Theater? How would the graphs of the functions compare?

*5. Drones are being launched from various platforms suspended in the air. The function f represents the height (in feet) of the first drone above the ground in terms of the number of seconds since the drone was launched, t. The functions $g, j,$ and k (defined in the items below) represent a second drone's possible height above the ground (in feet) in terms of the number of seconds since the *first* drone was launched, t.

For each given mathematical statement, do the following:
   i. Describe the second drone's launch and flight in comparison to the first drone's launch and flight.
   ii. Describe how the graphs of the two functions compare.

a. $g(t) = f(t) - 14$

b. $j(t) = f(t - 7)$

c. $k(t) = f(t - 2) + 38$

6. The graph of h is given. Draw the graph of g if $g(x) = -h(x+3)$.

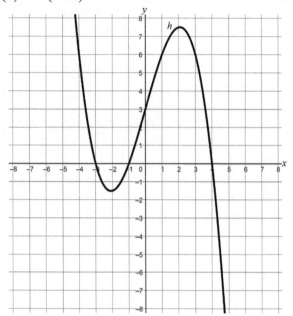

Many of the functions we have worked with in this module (and that we will continue to work with in the next few investigations) are ***polynomial functions***.

Polynomial Functions

When expressed as a formula, a polynomial function can be written in the form
$f(x) = a_n x^n + a_{n-1} x^{n-1} + a_{n-2} x^{n-2} + ... + a_2 x^2 + a_1 x + a_0$ where n, $n-1$, etc., are natural numbers
and a_n, a_{n-1}, etc., are real numbers. *Note that the coefficients could have a value of zero.*

Examples of polynomial function formulas include $f(x) = 4x^3 + 8.2x^2 - 3x + 1.7$ and
$g(x) = -5x^6 + 13x^2$. Quadratic functions and linear functions are examples of polynomials.

Quadratic = any polynomial where the biggest exponent
is 2 (i.e. x^2)

7. You are given the graphs of two polynomial functions that can be mapped onto each other using transformations. For each pair of polynomial functions, do the following.

    i.   Express the outputs of *g* in terms of the outputs of *f*. [*Your answer will look something like* $g(x) = f(x) + 7$ *or* $g(x) = f(2x)$.]

    ii. Express the outputs of *f* in terms of the outputs of *g*. [*Your answer will look something like* $f(x) = -g(x)$ *or* $f(x) = g(x + 3) - 4$.]

a.

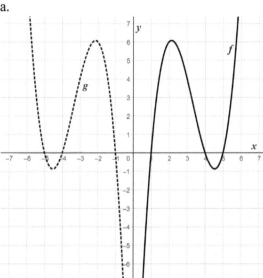

b.

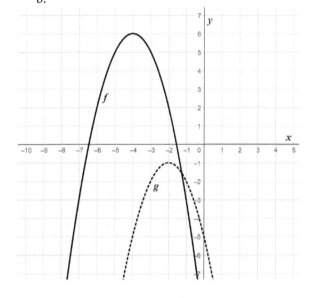

c.

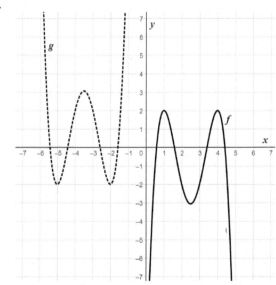

*1. Use your graphing calculator to sketch a graph of each of the following quadratic functions and compare their behaviors and properties. For each, determine any real-number roots for the function. (A ***root*** is a value of the independent variable that produces a value of 0 for the dependent variable. For example, if $f(3) = 0$, then 3 is a real-number root of function f.)

[handwritten: root = x value that produces 0 for y]

a. $f(x) = x^2$

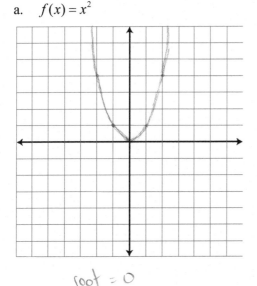

[handwritten: root = 0]

b. $g(x) = -3x^2 + 27$

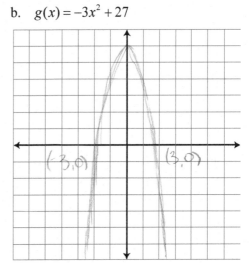

[handwritten: (-3,0)   (3,0)]

c. $h(x) = (x-2)(x+4)$

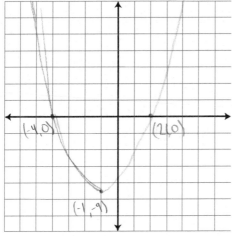

[handwritten: (-4,0)   (2,0)   (-1,-9)]

[handwritten: root = 2, -4]

d. $k(x) = -x^2 + 3x - 6$

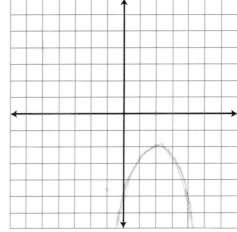

[handwritten: No real root]

*2. Complete each of the following. *It might help you to look back at your work in Exercise #1.*

a. When a quadratic function *f* is given in factored form, explain why the roots occur where each factor has a value of 0. [*Note: If function f is given in factored form, such as* $f(x) = (x-3)(x+5)$, *the factors are* $x-3$ *and* $x+5$.]

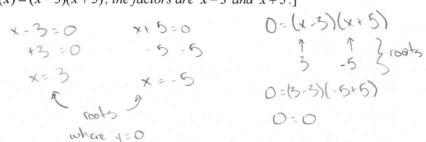

b. Is it possible for a quadratic function to have positive concavity over one interval of its domain and negative concavity over some other interval of its domain? Explain.

No. Always one or the other

c. Does a quadratic function always have a maximum value?

No. Always a maximum OR a minimum

d. How do you determine if a quadratic function will have positive or negative concavity?

Positive x = positive

Negative x = negative

e. What is the relationship between the *x*-coordinate where the maximum/minimum function value occurs and the function's roots?

x-coordinate is in the middle of roots
in between

*3. The function f is defined by $f(x) = (x+7)(x-3)$.

    a. Use algebraic methods to determine the roots of f.

$$-7 \text{ and } 3$$
$$(-7+7)(3-3) = 0$$

    b. Determine the x-coordinate of f's minimum value (also referenced as the x-coordinate of the vertex of the parabola generated by graphing f). Then find the function's minimum value.

$$x\text{-coordinate} = -2$$

$$f(2) = (-2+7)(-2-3)$$

    c. Evaluate $f(3)$.

$$f(3) = (3+7)(3-3)$$
$$f(3) = 0$$

It's relatively easy to determine the vertex and real roots (if any) of a quadratic function given in factored form. The roots are the values of x that make one of the factors 0, and the x-value of the vertex is halfway between the roots.

When the function is instead in **standard form** (in the form $f(x) = ax^2 + bx + c$ where a, b, and c are real numbers), the **quadratic formula** can be used to determine the function's roots.

The Quadratic Formula

For a quadratic function of the form $f(x) = ax^2 + bx + c$ where a, b, and c are real numbers*, the **quadratic formula** represents the function's roots (the values of x such that $f(x) = 0$).

$$x = \frac{-b}{2a} \pm \frac{\sqrt{b^2 - 4ac}}{2a}$$

*Note that b or c could be 0.

---

4. Use the quadratic formula to determine the roots of each given function (find both the exact values and decimal approximations). Then use a graphing calculator to check your work. *It might help to first list the values of a, b, and c to use in the formula.*

$$x = \frac{-b}{2a} \pm \frac{\sqrt{b^2 - 4ac}}{2a}$$

*a. $f(x) = 2x^2 + 5x + 1$

b. $f(x) = x^2 - 7x + 5$

*5. a. What does $x = \frac{-b}{2a}$ represent in the context of a quadratic function f?

The x-coordinate of the vertex

b. What does $f\left(\frac{-b}{2a}\right)$ represent in the context of a quadratic function f?

The y-coordinate of the vertex

c. What does $\pm\frac{\sqrt{b^2-4ac}}{2a}$ represent in the context of a quadratic function f?

How much you move left or right to find the roots
(distance from line of symmetry to roots)

d. If $\frac{\sqrt{b^2-4ac}}{2a} = 0$ for some quadratic function f, what does this tell us about the function?

Vertex is (1,0)

there are no roots on the x-axis

e. Assume that, for a given quadratic function f, the solutions to the quadratic formula are $x = -4 \pm 3.25$. Create a possible sketch of this quadratic function and illustrate the vertex and roots of this quadratic function.

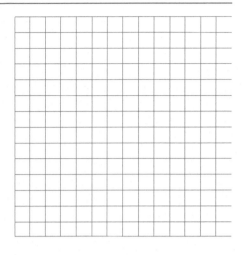

*6. Frozen Treats (an ice cream shop) finds that its weekly profit (in dollars), P, as a function of the price it charges per ice cream cone (in dollars), x, is modeled by
   $k(x) = -350x^2 + 3700x - 7480$ where $P = k(x)$.

   a. Determine the maximum weekly profit and the price of an ice cream cone that produces that maximum profit based on the model.

   b. If the cost of the ice cream cone is too low, then Frozen Treats will not make a profit. Determine what Frozen Treats needs to charge to break even (make a profit of $0.00).

   c. If the cost of the ice cream cone is too high, then not enough people will want to buy ice cream. As a result, the weekly profit will be $0.00. Determine what Frozen Treats would have to charge for this to happen.

d. The profit function for Cold & Creamy (another ice cream shop) is defined by the function g, where $g(x) = k(x-2)$. Does the function g have the same maximum value as k? What is the price per ice cream cone that Cold & Creamy ice cream shop must charge to produce a maximum weekly profit? Explain.

e. Describe the meaning of $h(x) = k(x) - 125$ and then compare/contrast the maximum values for each function.

7. Define the function formula that generates a parabola with horizontal intercepts at (–5, 0) and (3, 0) and passes through the point (2, 3).

*1. The graph of the function *g* represents the distance (in meters) of a diver's head above the water in terms of the number of seconds since the diver left a diving platform, *t*.

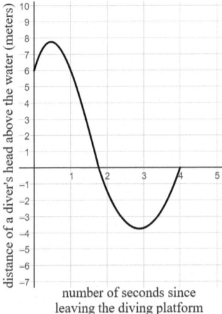

distance of a diver's head above the water (meters)

number of seconds since
leaving the diving platform

a.   What is the value of *g*(0)? What does this value represent in this context?

b.   What are the function's horizontal intercepts? What do they represent in this context?

c.   What does the graph of *g* convey about the diver's head's distance above the water during the first second since leaving the diving platform (0 ≤ *t* ≤ 1)?

d.   For what values of *t* is the diver's head below the water?

e.   On what interval(s) of *g*'s domain is the graph of *g* concave down?

*2. The function *f* is defined by $f(x) = (x+1)(x-3)^2$.

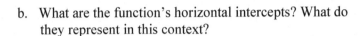

a.   Use algebraic methods to find the real-number roots (*x*-intercepts) of *f*.

-1 , 3

b.   What do a polynomial function's real-number roots represent?

where graph crosses x - axis

c. Explain why a polynomial function's roots occur where one of the factors has a value of zero.

$y = 0$ where roots are, one of the factors HAS to equal 0

d. Draw a number line representing values of x and highlight each of the following:

i. the intervals of x where f 's output is positive

ii. the intervals of x where f 's output is negative

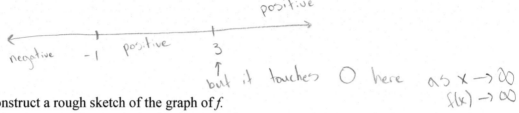

positive

negative -1 positive 3

but it touches 0 here   as $x \to \infty$
$f(x) \to \infty$

e. Construct a rough sketch of the graph of f.

as $x \to -\infty$
$f(x) \to -\infty$

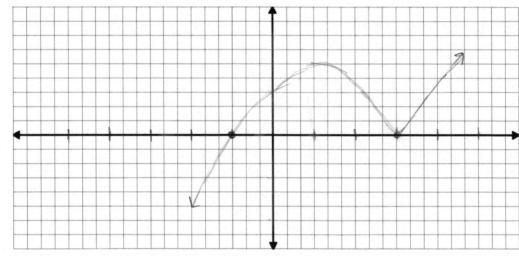

3. How do the functions f and h compare, given that $f(x) = (x+1)(x-3)^2$ and $h(x) = 2(x+1)(x-3)^2$?

4. How do the functions h and k compare, given that $h(x) = 2(x+1)(x-3)^2$ and $k(x) = -2(x+1)(x-3)^2$?

*5. Let $f(x) = x^3$, $g(x) = x^5$, $h(x) = x^2$, and $j(x) = x^8$. (A function of the form $p(x) = ax^n$, given that a and n are real numbers, is called a **power function**. Note that n must be a non-negative integer for p to be a polynomial function.)

a. Describe how each power function varies as: *"end behavior"*
    i. x increases without bound (also written "as $x \to \infty$"):

    As $x \to \infty$, $f(x) \to$ ___∞___   As $x \to \infty$, $g(x) \to$ ___∞___

    As $x \to \infty$, $h(x) \to$ ___∞___   As $x \to \infty$, $j(x) \to$ ___∞___

    ii. x decreases without bound (also written $x \to -\infty$):

    As $x \to -\infty$, $f(x) \to$ ___$-\infty$___   As $x \to -\infty$, $g(x) \to$ ___$-\infty$___

    As $x \to -\infty$, $h(x) \to$ ___∞___   As $x \to -\infty$, $j(x) \to$ ___∞___

b. What general statements can you make about how the exponent on a power function impacts the function's behavior?

Even exponent

as $x \to \infty$
$f(x) \to \infty$

as $x \to -\infty$
$f(x) \to \infty$

odd exponent

as $x \to \infty$
$f(x) \to \infty$

as $x \to -\infty$
$f(x) \to -\infty$

Reversed w/ negative coefficient

c. How does each function's behavior change if its coefficient becomes -2 instead of 1 (for example, if $f(x) = x^3$ changes to become $f(x) = -2x^3$)?

as $x \to \infty$
$f(x) \to -\infty$

as $x \to -\infty$
$f(x) \to \infty$

6. Given $f(x) = x^3 + 4x^2 - 6x - 12$ and $g(x) = x^3$, do the following.

   a. Graph the functions with a window size ranging from $x = -5$ to $x = 5$ and from $y = -20$ to $y = 20$. Do the functions have similar behaviors?

   b. Graph the functions with a window size ranging from $x = -50$ to $x = 50$ and from $y = -10,000$ to $y = 10,000$. Do the functions have similar behaviors?

   c. If your answers to parts (a) and (b) are different, why are they different?

A Polynomial Function's Leading Term

For a polynomial function f represented by a formula, the leading term is the term containing the largest exponent. For example, given $f(x) = 5x^4 + 7x^3 + 5x + 100$, the leading term is $5x^4$, and given $g(x) = 5 - 4x^3$, the leading term is $-4x^3$.

It's important to be able to identify the leading term for a polynomial function because the leading term dictates the function's **end behavior** (that is, the behavior of $f(x)$ as $x \to \pm\infty$, or as x increases or decreases without bound). This is because of the power of repeated multiplication with very large magnitude numbers – larger exponents create huge discrepancies in the relative sizes of each term in the long run.

For example, consider the function $f(x) = 5x^4 + 7x^3 + 5x + 100$. We can evaluate $f(10,000) = 50,007,000,000,050,100$, but how does each term contribute to this function value?

- $5(10,000)^4 = 50,000,000,000,000,000$, which is 99.986% of the value of $f(10,000)$.
- $7(10,000)^3 = 7,000,000,000,000$, which is 0.014% of the value of $f(10,000)$.
- $5(10,000) = 50,000$, which is 0.0000000001% of the value of $f(10,000)$.
- 100 is 0.0000000000002% of the value of $f(10,000)$.

As x continues to increase without bound, the value of $f(x)$ becomes virtually indistinguishable from the value of its leading term, $5x^4$. Thus, the value (and behavior) of $f(x)$ can be well-estimated by the value of $5x^4$ as $x \to \pm\infty$.

---

7. For each polynomial function, identify the leading term (the term with the largest exponent).

   a. $h(x) = -4x^3 + x^2 - 900x + 5$ *b. $f(x) = 9x^2 - 4x - 2x^4$ *c. $g(x) = -7(x-5)(x^2+1)$

   $-4x^3$ $-2x^4$ $-7(x^3-5+x-5x^2)$

                                                             $-7x^3$

Since a polynomial function behaves like its leading term in the long run, we can use notation to communicate this idea. Returning to $f(x) = 5x^4 + 7x^3 + 5x + 100$, we might write "As $x \to \pm\infty$, $f(x) \to 5x^4$." (We might read this as follows: "As x increases or decreases without bound, the value of $f(x)$ approaches the value of $5x^4$" or "As x increases or decreases without bound, the value of $f(x)$ can be well-estimated by the value of $5x^4$.")

Since we know (or can easily determine) the end behavior of $y = 5x^4$, we can thus predict the end behavior of f. As $x \to \infty$, $5x^4 \to \infty$, and as $x \to -\infty$, $5x^4 \to \infty$. Thus, this is also the end behavior of f. So, we can say two things.

- As $x \to \pm\infty$, $f(x) \to 5x^4$.
- As $x \to \infty$, $f(x) \to \infty$, and as $x \to -\infty$, $f(x) \to \infty$.

---

8. Using the same functions from Exercise #7, complete the following statements by first filling in the variable expression that approximates the function values as x increases or decreases without bound and then filling in the end behavior. *Part (a) is done for you.*

   a. $h(x) = -4x^3 + x^2 - 900x + 5$ *b. $f(x) = 9x^2 - 4x - 2x^4$ *c. $g(x) = -7(x-5)(x^2+1)$

   As $x \to \pm\infty$, $h(x) \to -4x^3$.   As $x \to \pm\infty$, $f(x) \to \underline{-2x^4}$.   As $x \to \pm\infty$, $g(x) \to \underline{-7x^3}$.

   As $x \to \infty$, $h(x) \to -\infty$.   As $x \to \infty$, $f(x) \to \underline{-\infty}$.   As $x \to \infty$, $g(x) \to \underline{-\infty}$.

   As $x \to -\infty$, $h(x) \to \infty$.   As $x \to -\infty$, $f(x) \to \underline{-\infty}$.   As $x \to -\infty$, $g(x) \to \underline{\infty}$.

9. Describe the end behavior of each of the following polynomial functions without first graphing them. Then check your work by graphing the functions with a graphing calculator or graphing software. *Remember to consider the role of the leading coefficient when determining a function's end behavior.*

   *a. $f(x) = -x^3 + 4x^2 - 8$ b. $g(x) = 5x - x^4 + 3x^5$ *c.
        $h(x) = -2(x+1)(x-3)$

*10. Answer the following questions given the graph of the polynomial function *g*.

a. What are the roots of *g*?

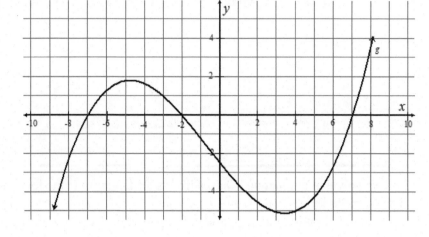

b. Evaluate $g(0)$.

c. On what interval(s) of the domain is the function increasing?

d. On what interval(s) of the domain is the function decreasing?

e. Estimate the location(s) of any inflection points.

f. On what interval(s) of the domain does the function have positive concavity?

g. On what interval(s) of the domain does the function have negative concavity?

h. Describe the function's end behavior.

i. When *g* is written as a formula of *y* in terms of *x*, is the largest exponent an even number or an odd number? How do you know?

The Number *i*

Have you ever tried to explain negative numbers to a child? For children, numbers are very concrete concepts – they use numbers to count items and (if they're old enough) perhaps to measure something like an object's height (in some units). In these examples negative numbers don't make much sense since they aren't needed for counting and children don't yet have a concept of directional measurement.

Believe it or not, mathematicians didn't always accept negative numbers. Until about 500 years ago almost every mathematician used geometry to justify their work, and characteristics such as length, area, and volume are represented by positive numbers. So mathematicians threw out any solutions to equations that were negative and believed negative numbers were useless. In this investigation we're going to explore another type of number that comes up when solving quadratic equations. It might be initially strange to think about these numbers, but they have very powerful applications in fields such as science and engineering.

1. Find the solutions to each of the following equations.
   a. $x^2 - 144 = 0$
   b. $x^2 - 49 = 0$

The equations $x^2 - 144 = 0$ and $x^2 - 49 = 0$ each have two real number solutions (one positive and one negative). But the equation $x^2 + 1 = 0$ does not have any real number solutions.

$$x^2 + 1 = 0$$
$$x^2 = -1$$
$$x = ????$$

Initially mathematicians ignored equations such as these because "they do not have a solution." However, eventually some mathematicians defined a new type of number to represent their solutions. They defined the number *i* to a number such that $i^2 = -1$.

If that is the case, then it must be that

$$i^2 = -1$$
$$i = \sqrt{-1}$$

The Number *i*

i is a number such that $i^2 = -1$. Then $i = \sqrt{-1}$, and $x = \pm i$ are the solutions to $x^2 + 1 = 0$.

Based on this idea, we can represent expressions such as $\sqrt{-25}$, $\sqrt{-19}$, and $-\sqrt{-2}$ in terms of this new number.

$\sqrt{-25}$	$\sqrt{-19}$	$-\sqrt{-2}$
$\sqrt{25 \cdot -1}$	$\sqrt{19 \cdot -1}$	$-\sqrt{2 \cdot -1}$
$\sqrt{25} \cdot \sqrt{-1}$	$\sqrt{19} \cdot \sqrt{-1}$	$-\sqrt{2} \cdot \sqrt{-1}$
$5i$	$19i$	$-\sqrt{2}i$

2. Rewrite each of the following expressions in terms of i.

    a. $\sqrt{-4}$ b. $\sqrt{-121}$ c. $\sqrt{-41}$ d. $\sqrt{-13}$

In Exercises #3-5, write the solutions to each equation in terms of the number i. Check your answers by substituting the values back into the original equation.

3. $x^2 + 9 = 0$ 4. $x^2 + 36 = 0$ 5. $x^2 + 59 = 0$

Complex Numbers and Arithmetic with the Number i

Since mathematicians initially rejected negative numbers, just imagine what some of them thought about the *square root* of a negative number! They thought such "numbers" had no place in serious mathematics. The famous mathematician René Descartes coined the term ***imaginary numbers*** to express his dislike. The name stuck even though we now know that these so-called "imaginary" numbers are actually very useful. It took mathematicians a long time to accept them, but now they are a critical part of modern mathematics.

Expressions involving "imaginary" numbers are called ***complex numbers***.

Complex Numbers

A ***complex number*** is a number of the form $a + bi$ made up of both a real number portion (a) and an imaginary number portion (bi).
- If $a = 0$, then the number is called *purely imaginary*.
- If $b = 0$, then the number is a *real number*.

Examples: $3 + 2i$ $5 - 7i$ $7i$ -14

Is it possible to have an expression involving the number i that cannot be written in the form $a + bi$? For example, are $i^0 - 6$, $10i^2$, $4 + i^3$, and $11i + 6i^4$ complex numbers? They each have i raised to an exponent other than 1. However, since we know that $i^2 = -1$ (and $i^0 = 1$ by definition) we can rewrite each expression as a complex number of the form $a + bi$.

$i^0 - 6$	$10i^2$	$4 + i^3$	$11i + 6i^4$
$1 - 6$	$10(-1)$	$4 + i^2 \cdot i$	$11i + 6(i^2)^2$
-5	-10	$4 + (-1) \cdot i$	$11i + 6(-1)^2$
$-5 + 0i$	$-10 + 0i$	$4 - i$	$11i + 6(1)$
			$6 + 11i$

In Exercises #5-13, rewrite each expression as a complex number of the form $a + bi$.

5. $2 + i^3$ 6. $10i^4$ 7. $2i^3 - 8i$ 8. $6i^5 + 4i$ 9. $i^6 - 4i$

10. $5i^{21} - 6i^2$ 11. $6i^{46} + 5i^3$ 12. $i(i^3 - 6i^2)$ 13. $(i^3)^2 + 2i^4 + 4i^3$

The number i follows most of the same basic properties as real numbers. For example, the rules of exponents work the same, as do the commutative and associative properties of addition and multiplication. Therefore we can perform arithmetic with complex numbers and simplify the results.

$$(3 + 5i) + (6 - 13i)$$
$$3 + 5i + 6 - 13i$$
$$3 + 6 + 5i - 13i$$
$$9 - 8i$$

$$(1 + 3i)(7 - i)$$
$$(1 + 3i)(7) + (1 + 3i)(-i)$$
$$(1)(7) + (3i)(7) + (1)(-i) + (3i)(-i)$$
$$7 + 21i - i - 3i^2$$
$$7 + 20i - 3i^2$$
$$7 + 20i - 3(-1)$$
$$7 + 20i + 3$$
$$10 + 20i$$

$$(4 - 3i)^2$$
$$(4 - 3i)(4 - 3i)$$
$$(4 - 3i)(4) + (4 - 3i)(-3i)$$
$$(4)(4) + (-3i)(4) + (4)(-3i) + (-3i)(-3i)$$
$$16 - 12i - 12i + 9i^2$$
$$16 - 24i + 9i^2$$
$$16 - 24i + 9(-1)$$
$$16 - 24i - 9$$
$$7 - 24i$$

Notice that when we perform operations with complex numbers we end up with complex numbers as the result.

In Exercises #14-24, rewrite each expression as a complex number of the form $a + bi$.

14. $2i(9 + 4i)$ 15. $(-4 + i) + (19 - 6i)$

16. $3(7-5i)-8(6+3i)$

17. $(1+4i)(5-2i)$

18. $9i+2i(7i^3+3i^2)$

19. $(7-6i)(3-i)$

20. $(3+5i)^2$

21. $(a+bi)(c+di)$

22. $4-\sqrt{-49}$

23. $-2\sqrt{-32}$

24. $\sqrt{3}\cdot\sqrt{-27}$

Geometric Interpretations of Complex Numbers

With real numbers, we know that multiplying by –1 creates a number's opposite, or a number with the same distance from 0 but on the opposite side (such as $4(-1) = -4$).

Thinking about this transformation as a 180° rotation centered at 0 yields an identical mapping of 4 to –4.

Since $i^2 = -1$, we can also interpret multiplication by i^2 as a 180° rotation centered at 0.

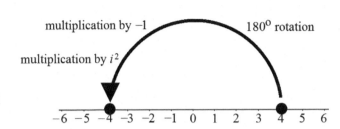

However, since multiplying by i^2 is identical to multiplication by i twice in succession ($4i^2 = 4 \cdot i \cdot i$), mathematicians realized that they could think about multiplication by i as a 90° rotation centered at 0 so that doing this twice in a row creates a 180° rotation.

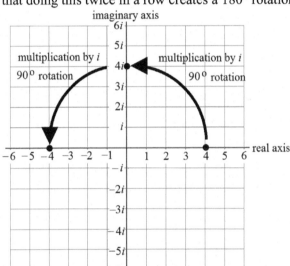

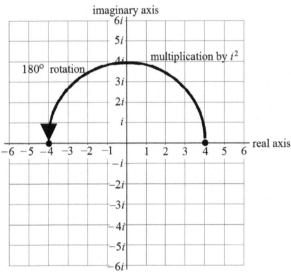

This way of thinking requires three additional ideas. 1) We can imagine that with any real number line there is a second perpendicular number line through 0 representing the imaginary numbers.
2) Multiplication by i is a 90° rotation counterclockwise (and multiplication by $-i$ is a 90° clockwise rotation).

Finally, 3) all complex numbers can be graphed on this ***complex plane*** with $a + bi$ represented by the point (a, b).

For example, on the complex plane we represent $4 + 2i$ by the point (4, 2). Multiplying by i yields

$$i(4 + 2i)$$
$$4i + 2i^2$$
$$4i + 2(-1)$$
$$-2 + 4i$$

Notice that this is a 90° counterclockwise rotation (centered at the intersection point of the axes) of the original point on the complex plane.

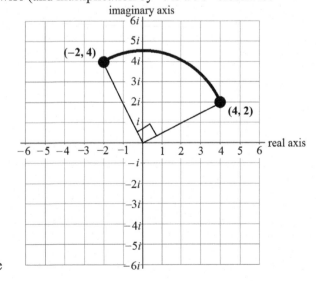

25. Represent each of the following complex numbers as points on the complex plane.

    a. $3+5i$

    b. $-6+2i$

    c. $-2-i$

    d. $7-4i$

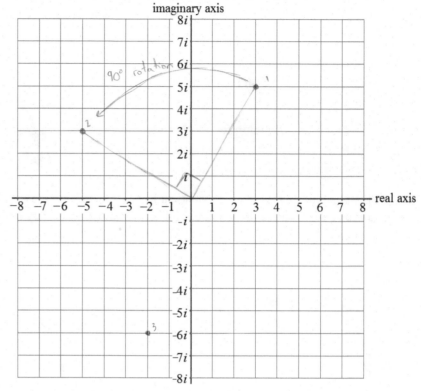

26. Rewrite each of the following products in the form $a + bi$ and plot them as points on the complex plane above. Verify that they are each 90° counterclockwise rotations of the original four points.

    a. $i(3+5i)$ b. $i(-6+2i)$ c. $i(-2-i)$ d. $i(7-4i)$

27 Rewrite each of the following products in the form $a + bi$ and plot them as points on the complex plane above. Verify that they are each 180° rotations of the original four points. *If your graph is getting crowded, you may complete this task on a piece of graph paper.*

    a. $i^2(3+5i)$ b. $i^2(-6+2i)$ c. $i^2(-2-i)$ d. $i^2(7-4i)$

28. Rewrite each of the following products in the form $a + bi$ and plot them as points on the complex plane above. Verify that they are each 90° clockwise rotations of the original four points. *If your graph is getting crowded, you may complete this task on a piece of graph paper.*

    a. $-i(3+5i)$ b. $-i(-6+2i)$ c. $-i(-2-i)$ d. $-i(7-4i)$

1. Use the quadratic formula to find the roots of $f(x) = -3x^2 + x - 4$. Comment on anything important about your solution.

$$x = -\frac{b}{2a} \pm \frac{\sqrt{b^2 - 4ac}}{2a}$$

$$\frac{-1}{-6} + \frac{\sqrt{1^2 - 4(-3)(-4)}}{2(-3)}$$

$$\frac{1}{6} \pm \frac{\sqrt{1 - 48}}{-6}$$

$$\frac{1}{6} \pm \frac{i\sqrt{47}}{-6}$$

2. The quadratic formula is supposed to find the zeros of the function. When the zeros are real numbers, they equate to the horizontal intercepts of the function's graph. However, in Exercise #1 the quadratic formula didn't return any real number answers. Graph function f using a calculator and explain what this fact implies about the horizontal intercepts of the graph of f.

When quadratic formula returns complex numbers, there aren't roots. No x-intercepts.

3. Use the quadratic formula to find the roots of $f(x) = 2x^2 - 4x + 2$. Comment on anything important about your solution.

4. The quadratic formula was used to find the zeros of each of the following functions.

$f(x) = x^2 + 7x + 25$

$g(x) = x^2 - 2x - 5$

$h(x) = 2x^2 + 20x + 50$

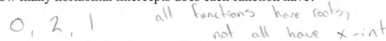

$x = -\frac{7}{2(1)} \pm \frac{\sqrt{(7)^2 - 4(1)(25)}}{2(1)}$

$x = -\frac{7}{2} \pm \frac{\sqrt{49-100}}{2}$

$x = -\frac{7}{2} \pm \frac{\sqrt{-51}}{2}$ → becomes imaginary

0 roots (2 complex roots)

$x = -\frac{(-2)}{2(1)} \pm \frac{\sqrt{(-2)^2 - 4(1)(-5)}}{2(1)}$

$x = -\frac{-2}{2} \pm \frac{\sqrt{4+20}}{2}$

$x = 1 \pm \frac{\sqrt{24}}{2}$

$x = 1 \pm \frac{2\sqrt{6}}{2}$

$x = 1 \pm \sqrt{6}$ → 2 roots, real numbers

$x = -\frac{20}{2(2)} \pm \frac{\sqrt{(20)^2 - 4(2)(50)}}{2(2)}$

$x = -\frac{20}{4} \pm \frac{\sqrt{400-400}}{4}$

$x = -5 \pm \frac{\sqrt{0}}{4}$

$x = -5 \pm 0$ → 1 root, right at (5,0)

How many horizontal intercepts does each function have?

0, 2, 1   all functions have roots, not all have x-int

5. The expression $b^2 - 4ac$ from the quadratic formula is called the **discriminant** because it discriminates (makes a distinction, or highlights the difference) between functions with zero, one, or two horizontal intercepts.

a. What is the value of the discriminant ($b^2 - 4ac$) for the functions from Exercise #4?

For $f(x)$: $b^2 - 4ac = -51$ negative = 0 x-int

For $g(x)$: $b^2 - 4ac = 24$ positive = 2 x-int

For $h(x)$: $b^2 - 4ac = 0$ 0 = 1 x-int

b. Why does the value of the discriminant give us enough information to predict the number of horizontal intercepts for a quadratic function?

The Discriminant of a Quadratic Function

Let f be defined by $f(x) = ax^2 + bx + c$ for real numbers a, b, and c. Then the discriminant is the value of

$b^2 - 4ac$

- If $b^2 - 4ac < 0$, then... the function won't have any x-intercepts

- If $b^2 - 4ac > 0$, then... the function will have two x-intercepts

- If $b^2 - 4ac = 0$, then... the function will have 1 ("double") x-intercept

For each of the functions in Exercises #6-9, use the discriminant to determine the number of horizontal intercepts for the function. [*You do not need to find the actual values.*] Graph the functions with a calculator or other graphing software to verify your answer.

6. $f(x) = 2x^2 + 4x + 2$

$b^2 - 4ac$

$4^2 - 4(2)(2)$

$16 - 16$

0

1 x-int

7. $f(x) = 2x^2 - 7x + 3$

$-7^2 - 4(2)(3)$

$49 - 24$

25

2 x-ints

8. $f(x) = x^2 + 3x + 5$

$3^2 - 4(1)(5)$

$9 - 20$

-11

0 x-int

9. $f(x) = 3x^2 - 12x + 12$

$-12^2 - 4(3)(12)$

$144 - 144$

0

1 x-int

When we know that the roots will be complex numbers, we can write them in the form $a + bi$. For example, consider the function $f(x) = x^2 + x + 2.5$. Let's verify that f has only complex roots and then find them. *Note that we will write the complex roots in the form $a + bi$.*

Solve the equation $f(x) = 0$, or $x^2 + x + 2.5 = 0$, for x.

$b^2 - 4ac$

$(1)^2 - 4(1)(2.5)$

$1 - 10$

-9

only complex solutions

$x = -\dfrac{1}{2(1)} \pm \dfrac{\sqrt{-9}}{2(1)}$

$= -\dfrac{1}{2} \pm \dfrac{3i}{2}$

$= -\dfrac{1}{2} \pm \dfrac{3}{2}i$

conjugate pair

$\boxed{x = -\tfrac{1}{2} + \tfrac{3}{2}i \text{ and } x = -\tfrac{1}{2} - \tfrac{3}{2}i}$

Remember that roots are input values that produce an output of 0. Let's demonstrate why $x = -\tfrac{1}{2} + \tfrac{3}{2}i$ and $x = -\tfrac{1}{2} - \tfrac{3}{2}i$ are roots of f.

$f\left(-\tfrac{1}{2} + \tfrac{3}{2}i\right) = \left(-\tfrac{1}{2} + \tfrac{3}{2}i\right)^2 + \left(-\tfrac{1}{2} + \tfrac{3}{2}i\right) + 2.5$

$= \left(\tfrac{1}{4} - \tfrac{3}{2}i + \tfrac{9}{4}i^2\right) + \left(-\tfrac{1}{2} + \tfrac{3}{2}i\right) + \tfrac{5}{2}$

$= \tfrac{1}{4} - \tfrac{3}{2}i + \tfrac{9}{4}i^2 - \tfrac{1}{2} + \tfrac{3}{2}i + \tfrac{5}{2}$

$= \tfrac{1}{4} + \tfrac{9}{4}(-1) - \tfrac{1}{2} + \tfrac{5}{2}$

$= \tfrac{1}{4} - \tfrac{9}{4} - \tfrac{1}{2} + \tfrac{5}{2}$

$= \tfrac{1}{4} - \tfrac{9}{4} - \tfrac{2}{4} + \tfrac{10}{4}$

$= 0$

$f\left(-\tfrac{1}{2} - \tfrac{3}{2}i\right) = \left(-\tfrac{1}{2} - \tfrac{3}{2}i\right)^2 + \left(-\tfrac{1}{2} - \tfrac{3}{2}i\right) + 2.5$

$= \left(\tfrac{1}{4} + \tfrac{3}{2}i + \tfrac{9}{4}i^2\right) + \left(-\tfrac{1}{2} - \tfrac{3}{2}i\right) + \tfrac{5}{2}$

$= \tfrac{1}{4} + \tfrac{3}{2}i + \tfrac{9}{4}i^2 - \tfrac{1}{2} - \tfrac{3}{2}i + \tfrac{5}{2}$

$= \tfrac{1}{4} + \tfrac{9}{4}(-1) - \tfrac{1}{2} + \tfrac{5}{2}$

$= \tfrac{1}{4} - \tfrac{9}{4} - \tfrac{1}{2} + \tfrac{5}{2}$

$= \tfrac{1}{4} - \tfrac{9}{4} - \tfrac{2}{4} + \tfrac{10}{4}$

$= 0$

In Exercises #10-15, do the following.
  a) Use the discriminant to verify that the function has only complex roots.
  b) Determine the function's roots. Write your final answers in the form $a + bi$.

10. $h(x) = x^2 + 6x + 10$

11. $f(x) = 3x^2 - 9x + 7.5$

12. $f(x) = x^2 + 5x + 9$

13. $f(x) = 4x^2 + 2x + 3$

14. $f(x) = 2x^2 + 5x + 6$

15. $f(x) = -3x^2 + 3x - 4$

I. THE BOTTLE PROBLEM – MODELING CO-VARYING RELATIONSHIPS (TEXT: S2, 3, 4, 5)

1. a. A spherical fishbowl fills with water. Describe how the height of the water and volume of water in the bowl vary together for equal volumes of water added to the bowl.

   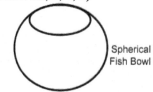
   Spherical Fish Bowl

   b. Construct a graph that represents the height of the water in the bowl as a function of the volume of water in the bowl.

2. The definition of an *increasing function* f follows: A function f is said to be increasing if $f(x_2) > f(x_1)$ whenever $x_2 > x_1$.

   a. Explain what this means in your own words. Use specific examples if it helps you.

   b. Given that g is a function defining the height of water in some bottle as a function of the volume of the water in the bottle, v, complete the following.
      i. Why must g be an increasing function?
      ii. Explain the meaning of the statement $g(v_2) > g(v_1)$ whenever $v_2 > v_1$ in this context.

   c. Use mathematical symbols to convey that another function, w, is <u>decreasing</u> for all values of x. How does $w(x)$ change as x increases?

3. You are given four graphs representing the heights of water in bottles as functions of the volumes of water in each bottle.
   i. Describe how the height of the water and volume of water in each bottle vary together for equal volumes of water added to the bottle.
   ii. Sketch a possible bottle that would generate each given graph. Include landmarks on both the bottle and graph to show points where the function's behavior changes in important ways.

   a.
   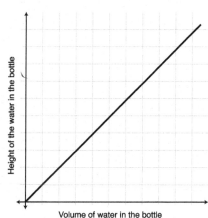

   b.
   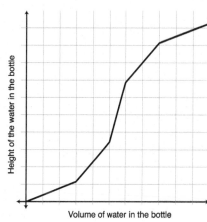

   c.
   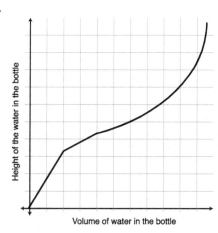

   d.
   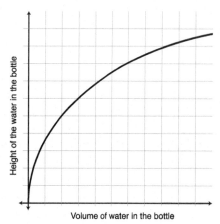

4. For each of the given scenarios, construct a graph of the height of the water in the bottle as a function of the volume of water in the bottle. Then, make an illustration of a bottle that would produce a height-volume graph with that general behavior.
   a. Scenario #1: A bottle in which the height of the water in the bottle increases at a constant rate with respect to the volume of water in the bottle.
   b. Scenario #2: A bottle in which the height of the water in the bottle initially increases at a constant rate with respect to the volume of water in the bottle, then switches so that equal changes in volume lead to smaller and smaller increases in the height of water in the bottle.
   c. Scenario #3: A bottle in which the height of the water in the bottle first increases by larger and larger amounts for equal increases in volume, then increases at a constant rate with respect to the volume of water in the bottle, and finally increases by smaller and smaller amounts for equal increases in volume.

5. As a runner moves around a quarter mile track, a radar gun detects the runner's direct distance from the starting line.
   a. Construct a graph that represents the direct distance (in yards) of the runner from the starting line in terms of the total distance (in yards) the runner has traveled around the track.
   b. At approximately what point(s) around the quarter mile track is the direct distance of the runner from the starting line at its maximum value?
   c. Provide a written description of how the runner's direct distance (in yards) from the starting line varies with the distance (in yards) the runner has traveled around the track.

6. Expand the following expressions as much as possible. *Combine like terms to simplify your answer.*
   a. $(x-3)(x+4)$ b. $-(x-7)^2$ c. $(x-4)^2(x-1)$ d. $3(x-2)^2$ e. $-(3x-2)(x+4)(2x-1)$

II. AVERAGE RATE OF CHANGE AND CONCAVITY (TEXT: S4, 5)

7. The values in the table below represent the distance (measured in yards) of a dog from a park entrance as a function of the number of seconds since the dog entered the park.

Change in the number of seconds since the dog entered the park	The number of seconds since the dog entered the park	The distance (in yards) of a dog from the park entrance	Change in the distance (in yards) of a dog from the park entrance	Average rate of change of the dog's distance with respect to time (yards per second)
	0	0		
	3	26.25		
	3.2	30.976		
	4.5	77.344		
	7	271.25		

   a. Complete the table of values.
   b. Describe what the average rates of change tell you about how the dog's distance from the park entrance changes over the time interval from $t = 0$ to $t = 7$ seconds.

8. When graphing the height of the water in a bottle with respect to the volume of water in the bottle, *inflection points* on the height–volume graph correspond to where the bottle changes from getting narrower to getting wider (or vice versa).

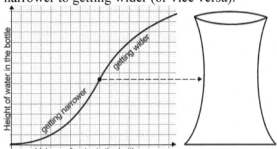 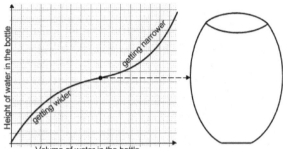

   a. Explain what the inflection point on the height-volume graph for the bottle on the left conveys about the changes in the water height and water volume as the bottle is being filled with water.

   b. Explain what the inflection point on the height-volume graph for the bottle on the right conveys about the changes in the water height and water volume as the bottle is being filled with water.

9. For the following functions, determine if the average rates of change of the water height with respect to the water volume over successive equal-sized intervals are increasing, decreasing, or constant.

   a.   b.   c.

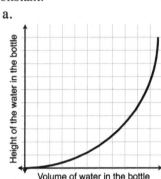 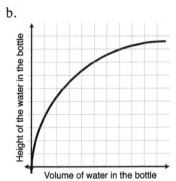 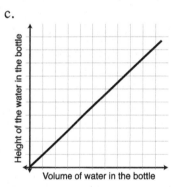

10. For each of the following functions, estimate the intervals on which:
        i.   the function values are increasing   ii. the function values are decreasing
       iii. the function has positive concavity   iv. the function has negative concavity

    a.   b.   c.

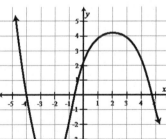 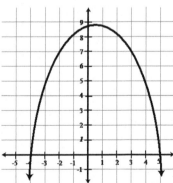

11. Using the functions in Exercise #10, determine the average rate of change of each function over the interval from $x = -2$ to $x = 3$. *You may need to estimate the function values to complete this task.*
    $x = -2$ to $x = 3$. *You may need to estimate the function values to complete this task.*

12. Fire pits are frequently used to cook meat. A fire is built to heat the rocks in the bottom of the pit. Once the pit reaches the desired temperature, the meat is then put in the pit and covered.

Number of hours since meat was added to the pit	Air temperature of the pit (in degrees Fahrenheit)
0	575
2	400
4	320
6	275

   The given table represents the air temperature (in degrees Fahrenheit) of a pit in terms of the number of hours that have elapsed since meat was added to it.

   a. Using the table of values, do the following.
      i. Sketch a graph of the pit's air temperature (in degrees Fahrenheit) in terms of the number of hours since the meat was added to the pit. Be sure to label your axes.
      ii. On your graph from part (i), represent the changes in the air temperature of the pit from 0 hours to 2 hours; from 2 hours to 4 hours; and from 4 hours to 6 hours.
      iii. What do you notice about the changes in the air temperature of the pit for successive equal changes in the number of hours since the meat was added to the pit?
   b. Let t represent the number of hours since the meat was added to the pit. Determine the average rate of change of the air temperature of the pit on each of the following time intervals, and then describe how to interpret these values.
      i. $0 \leq t \leq 2$ ii. $2 \leq t \leq 4$ iii. $4 \leq t \leq 6$
   c. Does this function have positive concavity, negative concavity, or some combination of both on the interval $0 < t < 6$?

13. $f(t) = 1.5^t$, with $d = f(t)$, represents a car's distance d (measured in feet) from a stop sign in terms of the number of seconds t since the car started to move away from the stop sign.
   a. Sketch a graph of this relationship. Be sure to label your axes.
   b. Determine the average rate of change of the distance of the car from the stop sign on each of the following time intervals, then explain how to interpret each value.
      i. $0 \leq t \leq 2$ ii. $2 \leq t \leq 4$ iii. $4 \leq t \leq 6$ iv. $1 \leq t \leq 5$
   c. Does this function have positive concavity, negative concavity, or some combination of both on the interval $0 < t < 6$?

14. The given graph represents Sally's distance from her house (measured in yards) in terms of the number of minutes since Sally started walking, t.
   a. Interpret the meaning of the point (20, 1950).
   b. Represent the changes in Sally's distance from her house (in yards) from 0 minutes to 10 minutes; from 10 minutes to 20 minutes; and from 20 minutes to 30 minutes.
   c. What do you notice about the change in Sally's distance from her house for successive equal changes in the number of minutes since Sally started walking?
   d. Describe what these successive changes in Sally's distance from her house for equal change in the number of minutes since Sally started walking tell you about how Sally's distance from her house is changing over the time interval from $t = 0$ to $t = 30$ minutes.
   e. Describe how Sally's speed (in yards per minute) is changing as the number minutes since Sally started walking increases from 0 to 30 minutes.

III. TRANSFORMATIONS OF POLYNOMIAL FUNCTIONS (TEXT: S9)

15. The graph of a polynomial function f is given. Sketch a graph of each of the following functions.

    a. $g(x) = -f(x)$ b. $h(x) = f(-x)$

    c. $k(x) = f(x-2)$ d. $p(x) = f(x) - 2$

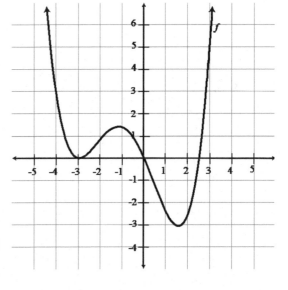

16. A coffee shop models its weekly profit P (measured in dollars) as a function of the price x (measured in dollars per cup) using
    $f(x) = -2000x^2 + 8000x - 2000$ with $P = f(x)$.

    a. A second coffee shop models its weekly profit as a function of the price it charges per cup, x, using function g where $g(x) = f(x-1)$. Discuss the relationship between the inputs and outputs of functions g and f.

    b. A third coffee shop models its weekly profit as a function of the price it charges per cup, x, using function h where $h(x) = f(x) - 150$. Discuss the relationship between the inputs and outputs of functions h and f.

    c. Do either g or h represent the same maximum profit as f? Explain.

17. The graphs of polynomial functions f and g are given.

    a. How do the input and output pairs for the two functions compare?

    b. Express f in terms of g.

    c. Express g in terms of f.

18. The graphs of polynomial functions f and g are given.

    a. How do the input and output pairs for the two functions compare?

    b. Express f in terms of g.

    c. Express g in terms of f.

    d. How does the average rate of change for each function compare on the interval $-1 < x < 3$?

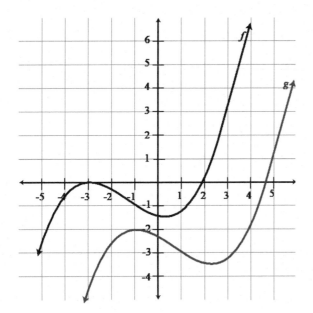

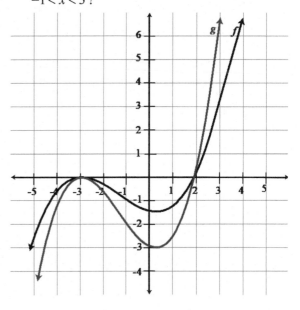

19. Each summer, Primo Pizza and Pizza Supreme compete to see who has the larger summer profit. Let f be the function that determines Primo Pizza's profits in terms of the number of days since June 1. Let g be the function that determines Pizza Supreme's profits in terms of the number of days since June 1.

    Let x represent the number of days since June 1.
    a. If Primo Pizza's profits each day are two times as large as Pizza Supreme's profits,
        i.   express the function f in terms of the function g.
        ii. express the function g in terms of the function f.
    b. If Primo Pizza's profits each day are two hundred dollars more than Pizza Supreme's profits,
        i.   express the function f in terms of the function g.
        ii. express the function g in terms of the function f.
    c. If Primo Pizza's profits on a given day are always the same as Pizza Supreme's profits two days later,
        i.   express the function f in terms of the function g.
        ii. express the function g in terms of the function f.

20. The function g is defined by $g(x) = 2f(x+3) - 4$. Explain how the behavior of function g compares to the behavior of function f.

21. Factor the following expressions as much as possible.
    a. $-8x^2 - 24x$
    b. $x^2 + 6x - 16$
    c. $2x^2 + 12x + 10$
    d. $2x^2 - 5x - 12$
    e. $3x^2 - 27$
    f. $6x^2 - x - 2$

IV. QUADRATIC FUNCTIONS (TEXT: S9)

22. Consider the functions $h(x) = x^2$ and $f(x) = 2^x$.
    a. How does the growth of the quadratic function h compare to the growth of the exponential function f?
    b. For large values of x, which of $f(x)$ or $h(x)$ will be larger?
    c. For what value(s) of x is the statement $f(x) = h(x)$ true?

23. Determine the roots of each quadratic function.
    a. $f(x) = (3x - 2)(x + 4)$
    b. $g(x) = x^2 - 9$
    c. $h(x) = 2x^2 + 5x + 3$
    d. $k(x) = -3x^2 + 14x - 16$

24. Determine the vertex of each quadratic function.
    a. $g(x) = (3x - 7)(2x + 6)$
    b. $j(x) = x^2 - 16$
    c. $p(x) = 4x^2 + 8x + 3$

25. Determine the roots and axis of symmetry of each quadratic function.
    a. $f(x) = x^2 - 6x + 9$
    b. $g(x) = (x + 3)(-2x - 5)$
    c. $h(x) = -x^2 + 7x - 6$

26. Let $f(x) = x^2$, $g(x) = (x - 4)^2$, $h(x) = (x - 4)^2 + 7$, and $p(x) = 3(x - 4)^2 + 7$.
    a. Determine the roots of each function.
    b. Determine the axis of symmetry for each function.
    c. Sketch a graph of each function.
    d. Compare the graphs (including roots, vertices, and output values) of f, g, h, and p.

27. Given a quadratic function g, the solutions to the quadratic equation are $x = 2 \pm 3.4$. Use this information to complete the following if possible. If not possible, say why.
    a. What are the roots of g?
    b. What is the vertex of g?

28. Function f represents the height of a ball above the ground, h (measured in feet), in terms of the amount of time since the ball was thrown upward from a bridge, t (measured in seconds). Then $f(t) = -16t^2 + 48t + 120$, with $h = f(t)$.
    a. Approximately how high is the bridge above the ground? Justify your answer.
    b. When does the ball hit the ground? Justify your answer.
    c. Construct a rough sketch of the graph of the function f that relates the height of the ball above the ground and the amount of time since the ball was thrown from the bridge.
    d. After how many seconds does the ball reach its maximum height above the ground? What is the maximum height above the ground reached by the ball? On the graph you drew in part (c), illustrate the point that represents the ball's maximum height above the ground.
    e. Pick two points on the graph and determine the function's average rate of change over the interval between them. Explain how to interpret this value.

29. A rock is thrown upward from a bridge that is 20 feet above a road. The rock reaches its maximum height above the road 0.91 seconds after it is thrown and contacts the road 2.35 seconds after it was thrown. Use your knowledge of the symmetry of a parabola and the given information to develop a quadratic function that relates the time since the rock was thrown to the height of the rock above the bridge.

30. A penny is thrown from the top of a 30.48-meter building and hits the ground 3.45 seconds after it was thrown. The penny reached its maximum height above the ground 0.823 seconds after it was thrown.
    a. Define a quadratic function h that expresses the height of the penny above the ground (measured in meters) as a function of the elapsed time since the penny was thrown (measured in seconds). (*Hint: Determine the zeros of the quadratic function, and then use the height of the building to find the value of the leading coefficient.*)
    b. What is the maximum height of the penny above the ground?

31. A coffee shop finds that its weekly profit, P (measured in dollars), is determined by the price x (in dollars) that the coffee shop charges for a cup of coffee. The profit P can be determined by the function $f(t) = -4000x^2 + 12000x - 4000$ with $P = f(t)$.
    a. Determine the maximum weekly profit and the price for a cup of coffee that produces the maximum profit.
    b. If $g(x) = f(x - 2)$, describe how the inputs and outputs of g and f are related. Does g have the same maximum profit as f? Explain.
    c. If $h(x) = f(x) - 2$, describe how the inputs and outputs of h and f are related. Does h have the same maximum profit as f? Explain.
    d. If $p(x) = f(x) + 60$, describe how the inputs and outputs of p and f are related.

32. Determine if each of the following are true or false. Provide a brief justification for your answers.
    a. A parabola must have either zero or two horizontal intercepts.
    b. If $f(x) = ax^2 + bx + c$ with $y = f(x)$, the vertical intercept of function is $(0, c)$.
    c. The vertex of a parabola is the point where the function values change from increasing to decreasing or from decreasing to increasing.
    d. The domain of a quadratic function (ignoring contexts) includes all real numbers.
    e. The range of a quadratic function (ignoring contexts) includes all real numbers.

f. If a quadratic function *f* has negative concavity on the interval $0 < x < 4$ then the values of $f(x)$ must decrease on this interval.

33. Use the quadratic formula to determine the exact roots of the following functions.
    a. $f(x) = x^2 + 5x$ b. $g(x) = x^2 + 4x - 3$ c. $k(x) = 10x^2 - 8x - 21$ d. $m(x) = -3x^2 - 2 + 12x$

34. Given each quadratic function in factored form, rewrite the formula in standard form ($ax^2 + bx + c$). Then construct a graph of each function and label its roots and maximum or minimum value.
    a. $h(x) = (2x)(3x - 4)$ b. $m(x) = (x - 5)(2x + 3)$ c. $f(x) = -(3x + 3)(2x - 4)$

35. Given each quadratic function in standard form, rewrite the formula in factored form. Then construct a graph of each function and label its roots and maximum or minimum value.
    a. $p(x) = 2x^2 - 5x - 3$ b. $h(x) = 3x^2 + 11x - 4$ c. $k(x) = -x^2 + 9x - 14$

36. a. Define the formula for a quadratic function that has horizontal intercepts at (–6, 0) and (2, 0) and a graph that passes through the point (0, 5).
    b. What is the parabola's vertex?

37. a. Define the formula for a quadratic function that has horizontal intercepts at (1, 0) and (3, 0) and a graph that passes through the point (0, –10).
    b. What is the parabola's vertex?

38. Use the quadratic formula to solve the following equations for *x*. (*Hint: First rewrite the equation so that it equals 0. This will allow you to utilize the quadratic formula.*)
    a. $-7x^2 + 13x = -6$ b. $3x^2 - 4x - 4 = 12$ c. $-6x^2 + 22 = 35x$ d. $2x^2 + 5x = 12x - 2$

39. Why are second degree polynomials called "quadratics"? *You may conduct research on the Internet to determine the answer.*

V. ROOTS AND END BEHAVIOR OF POLYNOMIAL FUNCTIONS (TEXT: S1, 6, 7, 8)

40. What is the general form of a polynomial function formula? Identify the constant term and leading coefficient.

41. Which of the following are polynomial functions. Justify your answer.
    a. $f(x) = 2^x$ b. $g(x) = 5$ c. $h(x) = \frac{5}{x} - 3x^2$ d. $p(x) = 5x^4 + 3x^2 - 122$ e. $r(x) = 5^{-2} - 3x$

42. Determine the roots of the following polynomial functions.
    a. $g(x) = 3x(2x - 4)(x + 2)^2$ b. $h(x) = x^2(x - 4)^2(x^3 - 8)$
    c. $j(x) = x^3 + 6x^2 + 3x$ d. $k(x) = 2x^2 - 5x - 3$

43. Determine the roots of the following polynomial functions.
    a. $f(x) = 4x^2 + 8x + 2$ b. $g(x) = x^3 - x^2$
    c. $s(x) = -(2x + 4)(3x - 1)^2(x - 2)$ d. $n(x) = 2x(x - 7)(3x + 4)$

44. Given the function $f(x) = (x - 3)(x + 1)(x - 2)$, complete the following.
    a. Determine the roots of *f*.
    b. As *x* increases without bound, describe the behavior of $f(x)$.
    c. As *x* decreases without bound, describe the behavior of $f(x)$.
    d. What is the behavior of *f* on the intervals $-1 < x < 2$ and $2 < x < 3$?
    e. Use your responses in parts (a) through (d) to sketch a graph of *f*.

45. For the following polynomial functions determine:
    i. the interval(s) on which the output of the function is positive.
    ii. the interval(s) on which the output of the function is negative.
    a. $f(x) = x(3x + 6)(x - 1)$ b. $g(x) = 2x^3 - 4x^2 + 4x$ c. $h(x) = 3x(2x - 5)(x + 1)$

46. Define three polynomial functions that have roots at $x = 2$, $x = 4$, and $x = -3$.

47. Define three polynomial functions that have roots at $x = 1$, $x = -1$, and $x = -5$.

48. Find the formula for a polynomial function that has only three roots at $x = 2$, $x = 5$, and $x = -4$ and passes through the point $(3, 6)$.

49. For each given polynomial function, identify the leading term and describe the end behavior of the function based on your analysis of the leading term.
    a. $f(x) = 12x^3 - 2x^5 + 3x - 2$ b. $g(x) = 3x(x - 2)(x + 4)(-2x + 3)^2$
    c. $h(x) = 3x^3 - 4x + 8x^{10} - 3$ d. $m(x) = -2(4 + x)(2x - 7)$

50. For each given polynomial function, identify the leading term and describe the end behavior of the function based on your analysis of the leading term.
    a. $f(x) = 3x(x - 7)(x + 2)(x - 4)$ b. $h(x) = 2x^2(-x + 4)(x - 7)^2$
    c. $p(x) = 2x(3x - 7)(4x + 1)$ d. $s(x) = 3x^2 + 4x - 7x^8 + 6x - 2$

51. Given the function $f(x) = x^2(3x - 4)(2x + 5)^3$, complete the following.
    a. Determine the roots of f.
    b. Describe whether the graph of f will cross through (that is, change signs from positive to negative or vice versa) or only "bounce off" the horizontal axis at each of the roots.
    c. Evaluate $f(0)$.
    d. Examine the leading term of f to determine the end-behavior of the function.
    e. Determine on what interval(s) of the domain $f(x)$ is positive.
    f. Determine on what interval(s) of the domain $f(x)$ is negative.
    g. Use your responses in parts (a) through (f) to sketch a graph of f.

52. Given the function $g(x) = x^3(2x - 4)^2(3x - 2)(-x + 1)^2$, complete the following.
    a. Determine the roots of g.
    b. Describe whether the graph of g will cross through (that is, change signs from positive to negative or vice versa) or only "bounce off" the horizontal axis at each of the roots.
    c. Evaluate $g(0)$.
    d. Examine the leading term of g to determine the end behavior of the function.
    e. Determine on what interval(s) of the domain $g(x)$ is positive.
    f. Determine on what interval(s) of the domain $g(x)$ is negative.
    g. Use your responses in parts (a) through (f) to sketch a graph of g.

53. Use the graph of *f* to complete each part.
    a. What is the domain of *f*?
    b. What is the range of *f*?
    c. Evaluate $f(0)$.
    d. What are the roots of *f*?
    e. On what interval(s) is $f(x)$ increasing?
    f. On what interval(s) is $f(x)$ decreasing?
    g. On what interval(s) does *f* have positive concavity?
    h. On what interval(s) does *f* have negative concavity?
    i. When written as a formula, is the largest exponent an even number or an odd number? Explain your reasoning.
    j. Describe the behavior of $f(x)$ as *x* increases without bound.
    k. Describe the behavior of $f(x)$ as *x* decreases without bound.
    l. Are there any roots with an even multiplicity (even exponent)?

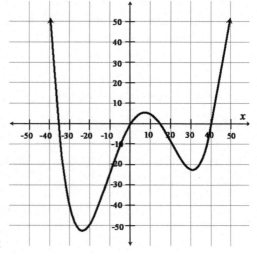

54. Use the graph of *g* to complete each part.
    a. What is the domain of *g*?
    b. What is the range of *g*?
    c. Evaluate $g(0)$.
    d. What are the roots of *g*?
    e. On what interval(s) is $g(x)$ increasing?
    f. On what interval(s) is $g(x)$ decreasing?
    g. On what interval(s) does *g* have positive concavity?

    h. On what interval(s) does *g* have negative concavity?

    i. When written as a formula, is the largest exponent an even number or an odd number? Explain your reasoning.
    j. Describe the behavior of $g(x)$ as *x* increases without bound.
    k. Describe the behavior of $g(x)$ as *x* decreases without bound.
    l. Are there any roots with an even multiplicity (even exponent)?

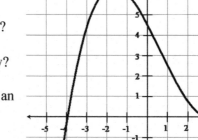

55. Evaluate each of the following.
    a. $f(x) = x(8.5 - 2x)(11 - 2x)$ when $x = 3.4$
    b. $g(x) = 87.4 - 53x$ when $x = 1.2$
    c. $h(x) = 3x^3 + 2x - 7$ when $x = -3.2$

56. Determine if each of the following are true or false. Provide a brief justification for your answers.
    a. The process of finding zeros of a polynomial function is also referred to as finding the roots of the polynomial function.
    b. We can apply the Zero-Product Property to determine the zeros of a polynomial function in factored form.
    c. The function used to model the box problem in Module 2 has exactly two roots at $x = 0$ and $x = 4.25$.
    d. The zeros of any function *f* represent the value(s) of *x* that, when input into the function *f*, return a value of 1 for the output $f(x)$.
    e. Polynomial functions are not always continuous.

INVESTIGATION 6: "IMAGINARY" NUMBERS

In Exercises #57-#68, rewrite each expression in terms of i.

57. $\sqrt{-9}$ 58. $\sqrt{-25}$ 59. $\sqrt{-16}$ 60. $\sqrt{-81}$ 61. $\sqrt{-23}$ 62. $\sqrt{-73}$

63. $-\sqrt{-144}$ 64. $\sqrt{-28}$ 65. $\sqrt{-80}$ 66. $-\sqrt{-24}$ 67. $\sqrt{-88}$ 68. $\sqrt{-60}$

In Exercises #69-72, write the solutions to each equation in terms of the number i. Check your answers by substituting the values back into the original equation.

69. $x^2 + 25 = 0$ 70. $x^2 + 121 = 0$ 71. $x^2 + 19 = 0$ 72. $x^2 + 23 = 0$

In Exercises #73-103, rewrite each expression as a complex number of the form $a + bi$.

73. $2 - \sqrt{-64}$ 74. $\sqrt{-144} + 5$ 75. $7 + \sqrt{-99}$ 76. $7\sqrt{-150} + 1$

77. $\sqrt{5} \cdot \sqrt{-20}$ 78. $7 + 2i^3$ 79. $3i^8$ 80. $10i^2 + 3i^5$

81. $9i^7 + i^3$ 82. $i^{16} + 5i^3$ 83. $3i^{53} - 8i^{22}$ 84. $i^{27} + i^{21}$

85. $i(i^5 + 2i^{10})$ 86. $i(4i^{15} + i^{27})$ 87. $(i^5)^3 - 8i^6 + 2i^4$ 88. $(4 + 6i) + (3 + 8i)$

89. $(8 - 2i) + (5 + i)$ 90. $(-6 + 8i) - (2 - 3i)$ 91. $\left(-\dfrac{6}{5} + 5i\right) + \left(\dfrac{9}{2} - \dfrac{16}{3}i\right)$

92. $(n + mi) - (v + ui)$ 93. $3i(2 + 3i)$ 94. $-2i(1 - i)$ 95. $(5 + i)(2 - 2i)$

96. $(-1 - i)(6 + 3i)$ 97. $(3 - 2i)^2$ 98. $(2 - 3i)^2$ 99. $(1 - 2i) - 2(3i)$

100. $(1 + \sqrt{2}i)(4 - \sqrt{3}i)$ 101. $(\sqrt{7} + 2i)(\sqrt{7} - 3i)$ 102. $i(6i + 4)^2$ 103. $di(p + ni)$

In Exercises #104-109, represent each complex number as a point on the complex plane [recall that $a + bi$ is represented by (a, b)]. *Perform the indicated multiplication and simplify the expression in the form $a + bi$ first if necessary.*

104. a. $2 + 5i$ b. $i(2 + 5i)$
     c. $i^2(2 + 5i)$ d. $-i(2 + 5i)$

105. a. $-4 + 6i$ b. $i(-4 + 6i)$
     c. $i^2(-4 + 6i)$ d. $-i(-4 + 6i)$

106. a. $-5 - 3i$ b. $i(-5 - 3i)$
     c. $i^2(-5 - 3i)$ d. $-i(-5 - 3i)$

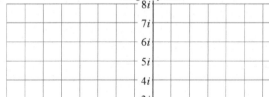

107. a. $1 - 7i$ b. $i(1 - 7i)$
     c. $i^2(1 - 7i)$ d. $-i(1 - 7i)$

108. a. $-6 + i$ b. $i(-6 + i)$
     c. $i^2(-6 + i)$ d. $-i(-6 + i)$

109. a. $-3 - i$ b. $i(-3 - i)$
     c. $i^2(-3 - i)$ d. $-i(-3 - i)$

110. Represent $4+2i$ and $3+3i$ as points on the complex plane. Then represent their sum $(4+2i)+(3+3i)$ as a point on the complex plane.

111. Represent $-3+4i$ and $5+i$ as points on the complex plane. Then represent their sum $(-3+4i)+(5+i)$ as a point on the complex plane.

112. Represent $-1-6i$ and $-3+4i$ as points on the complex plane. Then represent their sum $(-1-6i)+(-3+4i)$ as a point on the complex plane.

113. Represent $-2+5i$ and $4-2i$ as points on the complex plane. Then represent their sum $(-2+5i)+(4-2i)$ as a point on the complex plane.

114. How is the location of $(a+bi)+(c+di)$ on the complex plane related to the locations of $a+bi$ and $c+di$?

INVESTIGATION 7: COMPLEX SOLUTIONS TO THE QUADRATIC FORMULA AND THE DISCRIMINANT

In Exercises #115-123, use the discriminant to determine the number of horizontal intercepts for the function. [*You do not need to find the actual values.*] Graph the functions with a calculator or other graphing software to verify your answer.

115. $f(x) = 5x^2 + 3x - 1$ 116. $f(x) = 2x^2 + 10x + 12.5$ 117. $f(x) = 4x^2 - 14x + 12$

118. $f(x) = 4x^2 + 5x + 3$ 119. $f(x) = 6x^2 + x - 4$ 120. $f(x) = 2.25x^2 - 6x + 4$

121. $f(x) = 0.5x^2 + 4.5x + 10.125$ 122. $f(x) = -0.5x^2 - 3x - 6.125$ 123. $f(x) = -x^2 + x + 0.155$

In Exercises #224-232, do the following.
       a) Use the discriminant to verify that the function has only complex roots.
       b) Determine the function's roots. Write your final answers in the form $a + bi$.

124. $f(x) = x^2 - 2x + 2$ 125. $g(x) = x^2 - 4x + 29$ 126. $h(x) = x^2 + 8x + 25$

127. $f(x) = x^2 + 5x + 26.5$ 128. $h(x) = 2x^2 - 2x + 1$ 129. $f(x) = 3x^2 + 2x + 1$

130. $q(x) = 4x^2 + 53$ 131. $w(x) = -2x^2 - 3x - 2$ 132. $f(x) = -5x^2 + 6x - 7$

This investigation contains review and practice with important skills and procedures you may need in this module and future modules. Your instructor may assign this investigation as an introduction to the module or may ask you to complete select exercises "just in time" to help you when needed. Alternatively, you can complete these exercises on your own to help review important skills.

Simplifying Radicals

Use this section prior to the module or at any point as a spiraled review of skills.

One property of radicals is represented by the statement $\sqrt[n]{a \cdot b} = \sqrt[n]{a} \cdot \sqrt[n]{b}$. Basically, the nth root of a product is equivalent to finding the nth root of each factor and multiplying the results.

Three examples are shown that demonstrate this property.

$$\sqrt{64} \qquad\qquad \sqrt{100} \qquad\qquad \sqrt[3]{216}$$
$$\sqrt{16 \cdot 4} \qquad\qquad \sqrt{25 \cdot 4} \qquad\qquad \sqrt[3]{27 \cdot 8}$$
$$\sqrt{16} \cdot \sqrt{4} \qquad\qquad \sqrt{25} \cdot \sqrt{4} \qquad\qquad \sqrt[3]{27} \cdot \sqrt[3]{8}$$
$$4 \cdot 2 \qquad\qquad 5 \cdot 2 \qquad\qquad 3 \cdot 2$$
$$8 \qquad\qquad 10 \qquad\qquad 6$$

These examples don't demonstrate WHY you would use this property (for example, we already knew that $\sqrt{64} = 8$). This property is most commonly used to rewrite irrational numbers where the number inside the radical can be written as a product of the power of an integer and another number. See the given examples.

$$\sqrt{50} \qquad\qquad \sqrt{192} \qquad\qquad \sqrt[3]{56}$$
$$\sqrt{25 \cdot 2} \qquad\qquad \sqrt{64 \cdot 3} \qquad\qquad \sqrt[3]{8 \cdot 7}$$
$$\sqrt{25} \cdot \sqrt{2} \qquad\qquad \sqrt{64} \cdot \sqrt{3} \qquad\qquad \sqrt[3]{8} \cdot \sqrt[3]{7}$$
$$5\sqrt{2} \qquad\qquad 8\sqrt{3} \qquad\qquad 2\sqrt[3]{7}$$

Take a moment to verify that the expressions at the beginning and end of each process are equivalent. For example, use a calculator to verify that $\sqrt{192}$ and $8\sqrt{3}$ have the same decimal approximation.

In Exercises #1-10, rewrite each expression so that the magnitude of the integer inside the radical is as small as possible. *For example, rewrite $\sqrt{18}$ as $2\sqrt{3}$ or $\sqrt{252}$ as $6\sqrt{7}$.*

1. $\sqrt{75}$ 2. $\sqrt{28}$ 3. $\sqrt{48}$ 4. $\sqrt{108}$ 5. $\sqrt{45}$

6. $\sqrt{98}$ 7. $\sqrt[3]{54}$ 8. $\sqrt[3]{72}$ 9. $\sqrt[4]{48}$ 10. $\sqrt[4]{162}$

Rewriting Expressions

Use this section prior to the module or with/after Investigations 1 and 2.

In Exercises #11-16, use the distributive property to rewrite each expression. Simplify if possible.

11. $2x(x^2 - 6)$

12. $-x(10 - x^2)$

13. $4x(x - 3y)$

14. $x^3(2x^4 - \frac{1}{2}x)$

15. $-\frac{2}{3}x(6x + 10)$

16. $-\frac{7}{2}xy(8xy^2 - x^5y^3)$

In Exercises #17-28, rewrite each expression in expanded form. *For example, write* $(x + 2)(2x + 5)$ *in the form* $2x^2 + 9x + 10$.

17. $(x + 4)(x + 5)$

18. $(x - 3)(x + 6)$

19. $(x + 7)(x - 7)$

20. $(x + 10)(x - 8)$

21. $(x - 3)(x - 6)$

22. $(x - 10)(x + 10)$

23. $(x - 1)(x - 4)$

24. $(x + 6)^2$

25. $(x - 5)^2$

26. $(2x + 7)(x + 2)$

27. $(4x - 5)(4x + 5)$

28. $(2x + 3)(3x - 2)$

In the next set of exercises, we will factor expressions. Remember that it can sometimes be helpful to identify common factors. For example, factoring $2x^3 - 16x^2 + 30x$ is easier when we recognize that $2x$ is a common factor of all three terms.

$$2x^3 - 16x^2 + 30x$$
$$2x(x^2 - 8x + 15)$$
$$2x(x - 3)(x - 5)$$

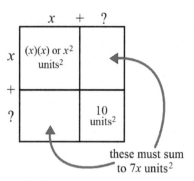

Also, recall the "area model" introduced in an earlier module. This model is often quite helpful for students who are trying to factor trinomial expressions like $x^2 + 7x + 10$. In this case the factored form is $(x + 2)(x + 5)$.

Finally, recall that sometimes a "middle term" can have a value of 0.

For example, if you want to factor the expression $x^2 - 25$ you might notice that there is no "x" term (only a term with x^2). However, without changing its value, we can rewrite the expression $x^2 - 25$ as $x^2 + 0 - 25$, or even better as $x^2 + 0x - 25$. The following examples demonstrate this idea.

$x^2 - 25$	$x^2 - 64$	$4x^2 - 9y^2$
$x^2 + 0x - 25$	$x^2 + 0x - 64$	$4x^2 + 0xy - 9y^2$
$(x + 5)(x - 5)$	$(x + 8)(x - 8)$	$(2x + 3y)(2x - 3y)$

29. a. Expand $(x - 6)(x + 6)$ to show that it is equivalent to $x^2 - 36$.

  b. Expand $(2x - 5)(2x + 5)$ to show that it is equivalent to $4x^2 - 25$.

In Exercises #30-41, factor each expression.

30. $x^2 + 16x + 60$

31. $x^2 - 5x - 24$

32. $x^2 - 100$

33. $5x^3 + 30x^2 + 40x$

34. $x^2 - 144$

35. $2x^2 - x - 28$

36. $4x^2 - 49$

37. $x^2 - 2xy + y^2$

38. $2x^3 + 5x^2 - 12x$

39. $16x^2 - 36$

40. $3a^2 + 6ab + 3b^2$

41. $6x^3 + 21x^2 + 15x$

Simplifying Rational Expressions
Use this section prior to the module or with/after Investigations 2 and 3.

Sometimes it's possible to factor out the same number from the numerator and denominator of a rational expression. This makes it possible to simplify the expression. For example, the numerator and denominator of $\dfrac{5x+35}{10}$ both have 5 as a common factor.

$$\frac{5x+35}{10}$$

$$\frac{5(x+7)}{5(2)}$$

Since $\frac{a}{b} \cdot \frac{c}{d} = \frac{a \cdot c}{b \cdot d}$ (which is true in "both directions"), we can rewrite this as follows.

$$\frac{5}{5} \cdot \frac{x+7}{2}$$

$$1 \cdot \frac{x+7}{2}$$

$$\frac{x+7}{2}$$

Thus, $\dfrac{5x+35}{10}$ can be simplified to $\dfrac{x+7}{2}$. They are equivalent rational expressions. We demonstrate two more examples.

$$\frac{12x-15}{3}$$
$$\frac{3(4x-5)}{3(1)}$$
$$\frac{3}{3} \cdot \frac{4x-5}{1}$$
$$1 \cdot \frac{4x-5}{1}$$
$$4x-5$$

$$\frac{14x+28}{21}$$
$$\frac{7(2x+4)}{7(3)}$$
$$\frac{7}{7} \cdot \frac{2x+4}{3}$$
$$1 \cdot \frac{2x+4}{3}$$
$$\frac{2x+4}{3}$$

In Exercises #42-47, simplify the rational expression if possible. If you can't simplify the expression, write "does not simplify".

42. $\dfrac{10x-20}{10}$

43. $\dfrac{8x-12}{9}$

44. $\dfrac{21x+3y}{3}$

45. $\dfrac{8x+40}{32}$

46. $\dfrac{35x-5y}{10}$

47. $\dfrac{18x-72}{45}$

It's also possible to simplify rational expressions when the numerator and denominator have the same variable factors.

$$\frac{2x^2+10x}{18x}$$

$$\frac{2x(x+5)}{2x(9)}$$

$$\frac{2x}{2x}\cdot\frac{x+5}{9}$$

$$1\cdot\frac{x+5}{9}$$

$$\frac{x+5}{9}$$

Continued on the next page.

HOWEVER, it's important to specify that $\dfrac{2x^2+10x}{18x}$ and $\dfrac{x+5}{9}$ are equivalent **only** as long as $x \neq 0$.

When $x = 0$, the original expression $\dfrac{2x^2+10x}{18x}$ is undefined. So, we say that $\dfrac{2x^2+10x}{18x} = \dfrac{x+5}{9}$ when $x \neq 0$. Two more examples follow.

$$\frac{x^2+6x+5}{x+5}$$

$$\frac{(x+5)(x+1)}{(x+5)(1)}$$

$$\frac{x+5}{x+5}\cdot\frac{x+1}{1}$$

$$1\cdot\frac{x+1}{1}$$

$$x+1 \quad \text{if } x \neq -5$$

$$\frac{x^2-7x+12}{x^2-2x-8}$$

$$\frac{(x-4)(x-3)}{(x-4)(x+2)}$$

$$\frac{x-4}{x-4}\cdot\frac{x-3}{x+2}$$

$$1\cdot\frac{x-3}{x+2}$$

$$\frac{x-3}{x+2} \quad \text{if } x \neq 4$$

Notice how we excluded $x = -5$ in the first example because it made the original expression undefined (but not the simplified form). We excluded $x = 4$ in the second example for the same reason.

In Exercises #48-53, simplify the rational expression if possible. If you simplify the rational expression, be sure to list x-values we must restrict. If you can't simplify the expression, write "does not simplify".

48. $\dfrac{x^2+5x-14}{x-2}$

49. $\dfrac{x^4+7x^2y}{3x^2}$

50. $\dfrac{x^2-8x-33}{x^2+5x+6}$

51. $\dfrac{x^2-7x+2}{x}$

52. $\dfrac{x+1}{3x^2-x-4}$

53. $\dfrac{x^3}{x^3-2x^2+5x}$

*1. Five gallons of liquid flavoring have been poured into a large vat. A water valve is opened, and water begins flowing into the vat to produce a mixed drink. Let x represent the volume of water in the mixture. Then the function f, defined by $f(x) = \frac{5}{x}$, represents the ratio of the volume of flavoring to the volume of water in the mixture.

Volume of flavoring (gallons)	Volume of water (gallons) x	$f(x) = \dfrac{5}{x}$
5	1/2	10
5	1	5
5	3	1.$\overline{6}$
5	5	1
5	8	0.625
5	10	0.5
5	50	0.1
5	500	0.01
5	2,000	0.0025

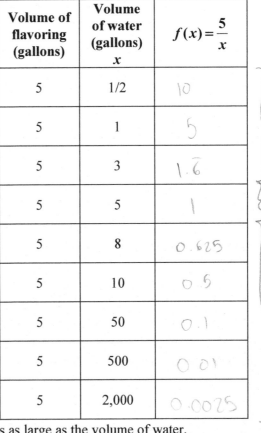

a. Complete the table of values showing the ratio of the volume of flavoring to the volume of water for each given value of x.

b. Use your completed table to answer the following.

   i. When 1 gallon of water has been added to the vat, the volume of flavoring is __5__ times as large as the volume of water.

   ii. When 3 gallons of water have been added to the vat, the volume of flavoring is __1.$\overline{6}$__ times as large as the volume of water.

   iii. When 5 gallons of water have been added to the vat, the volume of flavoring is __1__ times as large as the volume of water.

   iv. When 10 gallons of water have been added to the vat, the volume of flavoring is __0.5__ times as large as the volume of water.

   v. When 500 gallons of water have been added to the vat, the volume of flavoring is __0.01__ times as large as the volume of water.

   vi. When 2,000 gallons of water have been added to the vat, the volume of flavoring is __0.0025__ times as large as the volume of water.

   vii. As the number of gallons of water added to the vat increases without bound, the ratio of the volume of flavoring to the volume of water approaches __0__.

c. Label and scale the given axes and use your table from part (a) to plot at least 6 points of the form $(x, f(x))$. Then make a rough sketch of f's graph.

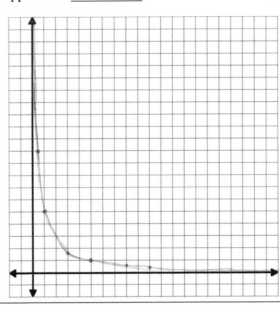

d. Complete the following statement for the graph you drew. [*This statement is asking what the value of the function f approaches as the value of x increases without bound (that is, as x becomes a larger and larger positive number).*]

As $x \to \infty$, $f(x) \to$ __0__.

(*Read: "As x increases without bound, $f(x)$ approaches __0__."*)

e. Where do you look on the graph of *f* to visualize the statement, "As $x \to \infty$, $f(x) \to$?"

function where output is ratio

polynomial ÷ polynomial the top is x times as large as the bottom

---

Rational Functions

If *p* and *r* are polynomial functions, then a function whose output values represent a ratio of the values of these functions (such as $f(x) = \frac{p(x)}{r(x)}$) is called a ***rational function***.

Recall that a polynomial function is a function that can be expressed in the form $p(x) = a_n x^n + a_{n-1} x^{n-1} + a_{n-2} x^{n-2} + \dots + a_1 x + a_0$ for some positive integer *n* and real numbers $a_0, \dots, a_n$.

Thus, a rational function can be expressed in the form

$$f(x) = \frac{p(x)}{r(x)} = \frac{a_m x^m + a_{m-1} x^{m-1} + a_{m-2} x^{m-2} + \dots + a_1 x + a_0}{b_n x^n + b_{n-1} x^{n-1} + b_{n-2} x^{n-2} + \dots + b_1 x + b_0}$$

for some positive integers *m* and *n* and real numbers $a_0, \dots, a_m$ and $b_0, \dots, b_n$.

When examining if a rational function's value approaches some constant as *x* increases without bound (as $x \to \infty$), we need to compare the relative sizes of the polynomial in the numerator ($p(x)$) and the denominator ($r(x)$) as we imagine the value of the input variable (*x*) getting larger and larger without bound.

Since a polynomial's end behavior (its value as $x \to \infty$) can be approximated by the polynomial's leading term (the term with the largest exponent), determining a rational function's end behavior involves comparing the leading term in the numerator and the leading term in the denominator, or the ratio $\frac{a_m x^m}{b_n x^n}$.

2. For each of the following functions:
    i. identify or determine the leading term (the term with the largest exponent) in the numerator and the leading term in the denominator,
    ii. complete the second column in the table by determining the relative size of the leading term in the numerator with the leading term in the denominator as x increases without bound,
    iii. complete the third column in the table by evaluating $h(x)$ at each value of x, and
    iv. describe the patterns that you notice by exploring the second and third columns in the table.

a. $h(x) = \dfrac{3x - 4}{2x^2 - 7x + 4}$

$\dfrac{3x}{2x^2} = \dfrac{3 \cdot x}{2 \cdot x \cdot x} = \dfrac{3}{2x}$

*b. $h(x) = \dfrac{x(x-3)(2x+4)}{5x^2 - 2x}$ *Hint: Before answering, rewrite the numerator in standard form by expanding the product* $x(x-3)(2x+4)$.

$\dfrac{2x^3}{5x^2} = \dfrac{2}{5}x$

x	$\dfrac{3x}{2x^2}$	$h(x)$
1		
2		
3		
10		
1000		
10,000		

x	$\dfrac{2x^3}{5x^2}$	$h(x)$
1		
2		
3		
10		
1000		
10,000		

c. $h(x) = \dfrac{4x^3 - 2x + 5000}{x(7x^2 + 20x - 9)}$ Hint:

Before answering, rewrite the denominator in standard form by expanding the product $x(7x^2 + 20x - 9)$.

x	$\dfrac{4x^3}{7x^3}$	$h(x)$
1		
2		
3		
10		
1000		
10,000		

Horizontal Asymptotes

A *horizontal asymptote* exists for the graph of *f* if a horizontal line $y = a$ can be used to estimate the end behavior of a function. If $f(x) \to a$ as $x \to \infty$ or $x \to -\infty$, then we say that $y = a$ is a horizontal asymptote for *f*.

Note that there are other possible end behaviors for functions (including rational functions). Not every rational function will have a horizontal asymptote. For example, if $f(x) \to -\infty$ as $x \to -\infty$ and if $f(x) \to \infty$ as $x \to \infty$, then the graph of f falls to the left and rises to the right and thus does not have a horizontal asymptote.

In Module 4, we explored the long-term behavior (or end behavior) of polynomial functions. We know that as *x* increases or decreases without bound, the behavior of a polynomial function is well-estimated by the behavior of its leading term.

For example, if $f(x) = x^2 - 10x + 5$, then as $x \to \pm\infty$ (as *x* increases or decreases without bound), *f* behaves similarly to $y = x^2$ (that is, using $y = x^2$ to estimate the value of $f(x)$ results in only a very small error as $x \to \pm\infty$). Thus, we might say the following.

$$\text{As } x \to \pm\infty, \ f(x) \to x^2 .$$

How might this help us predict the long-term behavior of rational functions? Let's explore.

*3. Given $f(x) = \dfrac{2x^2 - 6x + 1}{x - 5}$ (where $2x^2 - 6x + 1$ and $x - 5$ are polynomials), complete the following.

a. As x increases or decreases without bound (as $x \to \pm\infty$), what expression can we use to estimate the value of $2x^2 - 6x + 1$?

$$2x^2$$

b. As $x \to \pm\infty$, what expression can we use to estimate the value of $x - 5$?

$$x$$

c. As $x \to \pm\infty$, what expression do you think we can use to estimate the value of $f(x)$? [*Simplify the expression if possible.*]

$$\frac{2x^2}{x} = 2x$$

d. Using a graphing calculator or graphing software, graph function f along with $y = \#\#$ on the same set of axes, where "$\#\#$" is the expression you determined in part (c).

i. Zoom out on the graph. As you zoom out, what do you notice?

$$\text{as } x \to \infty, f(x) \to \infty$$
$$\text{as } x \to -\infty, f(x) \to -\infty$$

ii. What does your answer to part (i) tell us about the rational function f?

It behaves like $2x$

4. For each of the following rational functions, create an expression that estimates the function's value as $x \to \pm\infty$ and then state the function's horizontal asymptote (if one exists). Use a graphing calculator to check your work (see Exercise #3 part (d)).

*a. $f(x) = \dfrac{4x + 8}{2x - 9}$

b. $g(x) = \dfrac{x^3 + x - 20}{x + 4}$

c. $h(x) = \dfrac{x^2 - 6x + 5}{x^3 + 2}$

*d. $j(x) = \dfrac{(2x - 6)}{(3x - 4)(x + 2)}$

*5. Under what conditions does a rational function have:
   a. no horizontal asymptote? Explain and then give an example.

   b. a horizontal asymptote at $y = 0$? Explain and then give an example.

   c. a horizontal asymptote that is NOT $y = 0$? Explain and then give an example.

6. Given $f(x) = \dfrac{3x}{x+4}$, complete the following.
   a. Fill in the table of values.

x	$3x$	$x+4$	$f(x) = \dfrac{3x}{x+4}$
$-1{,}000{,}000$			
$-10{,}000$			
-100			
100			
$10{,}000$			
$1{,}000{,}000$			

b. As x increases without bound ($x \to \infty$), what happens to the value of $f(x)$? Why?

c. As x decreases without bound ($x \to -\infty$), what happens to the value of $f(x)$? Why?

d. Using a calculator, graph f. Explain how the graph supports your conclusions in parts (b) and (c) above.

7. A national park research team noticed a dramatic reduction in the deer population in a 150,000-acre protected area. To increase the deer population, the park services introduced 125 additional deer into the area. The researchers' population model predicts that the expected number of deer will follow the model $f(t) = \dfrac{50(2t + 16)}{0.0075t + 4}$, where t represents the number of years since the 125 deer were introduced.

   a. What was the deer population before the 125 deer were introduced?

   b. Find the population when i) $t = 7$ years; ii) $t = 18$ years; and iii) $t = 110$ years.

c. When will the population of deer reach 760?

d. What expression can be used to estimate the long-term behavior of the function's values?

e. The research team's model predicts that the total number of deer that can be supported by the 150,000-acre area is limited. What is the maximum number of deer that the research team expects the 150,000-acre area to support? Discuss your approach with your group or as a class.

8. For each given function, do the following.
   i.   Write a variable expression that can be used to estimate the function's value as $x \to \pm\infty$. Simplify the expression if possible. Use a form similar to, "As $x \to \pm\infty$, $f(x) \to$ _____."
   ii. State the function's horizontal asymptote (if one exists).
   iii. Verify your work by graphing the function using a graphing calculator or graphing software.

*a. $f(x) = \dfrac{8}{x+1}$

b. $g(x) = \dfrac{x}{2x-8}$

*c. $h(x) = \dfrac{6x^2 - 24}{2(x-3)(x+1)}$

d. $m(x) = \dfrac{x+3}{x^2 + 4x - 5}$

*e. $n(x) = \dfrac{x^5 + 1}{x^2 + 1}$

f. $c(x) = \dfrac{2(x+6)}{x+6}$

Function Outputs as Ratio Values

In Investigation #1, we explored functions with outputs representing the value of a ratio between two other quantities' values as they change. The output of a rational function doesn't tell us the values of these two quantities – it tells us the <u>relative size</u> of these quantities.

For example, if a vat begins with 5 gallons of liquid flavoring and x gallons of water are added, then $f(x) = \frac{x}{x+5}$ models the portion of the total mixture that is water in terms of x. Then $f(15) = 0.75$ doesn't tell us the volume of the water, flavoring, or total mixture. It tells us that the volume of water is 0.75 (or 75%) of the total mixture volume (whatever that volume is).

---

*1. Given that $f(x) = \dfrac{2}{x}$, what happens to the value of $f(x)$ as x gets close to 0? Use the given table to help you formulate your response.

As $x \to 0$

$f(x) \to \infty$ and $-\infty$

as $x \to 0^-$

$f(x) \to \infty$

as $x \to 0^+$

$f(x) \to \infty$

x	$f(x) = \frac{2}{x}$
−2	−1
−1	−2
−0.1	−20
−0.01	−200
−0.001	−2000 −∞
0	undefined
0.001	2000 ∞
0.01	200
0.1	20
1	2
2	1

"Approaches" Notation

When the values of x approach a constant (such as 0) from values less than that number (that is, the values are increasing toward 0), we say that "x approaches 0 from the left" and write this in notation as $x \to 0^-$ (*because the direction comes from the negative end of the number line*).

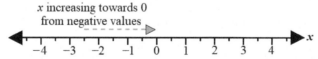

x increasing towards 0
from negative values

When the values of x approach a constant (such as 0) from values greater than that number (that is, the values are decreasing toward 0), we say that "x approaches 0 from the right" and write this in notation as $x \to 0^+$ (*because the direction comes from the positive end of the number line*).

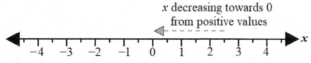

x decreasing towards 0
from positive values

---

2. For each of the following sequences of values for a variable x, use "approaches" notation to indicate what value x is approaching and the direction of change. (*Your answers will look like* $x \to 3^-$ *or* $x \to -9^+$.)

*a. 6, 5, 4.6, 4.55, 4.51, 4.5001, 4.500001, ...

$x \to 4.5^+$ $4.5 \Leftarrow 6$

b. 9, 9.9, 9.99, 9.999, 9.9999, ... $9 \to 10$

$x \to 10^-$

*c. $-6, -5.5, -5.1, -5.01, -5.001, ...$

$x \to -5^-$ $-6 \to -5$

d. $-1, -1.5, -1.9, -1.99, -1.999, ...$ $-2 \quad -1$

$x \to -2^+$

3. Determine whether the following sequences of numbers are increasing or decreasing. (*Note: A sequence of all negative numbers is increasing if successive numbers are getting closer to 0.*)

*a. $\dfrac{1}{2}, \dfrac{1}{4}, \dfrac{1}{8}, \dfrac{1}{16}, \dfrac{1}{32}$

b. $\dfrac{2}{100}, \dfrac{5}{40}, \dfrac{10}{35}, \dfrac{15}{9}, \dfrac{50}{6}$

*c. $-\dfrac{1}{80}, -\dfrac{1}{40}, -\dfrac{1}{20}, -\dfrac{1}{10}, -\dfrac{1}{5}$

d. $\dfrac{-1}{100}, \dfrac{-1}{200}, \dfrac{-1}{320}, \dfrac{-1}{450}, \dfrac{-1}{1089}$

*4. If $f(x) = \dfrac{1}{(x-8)^2}$, complete the following.

a. Complete the table of values. *Pay attention to the values of the numerator and denominator as you perform your calculations.*

x	$f(x) = \frac{1}{(x-8)^2}$
7	
7.5	
7.9	
7.99	
7.999	
8	undefined
8.001	
8.01	
8.1	
8.5	
9	

b. What do you notice as x approaches 8 from the left (from values less than 8, written $x \to 8^-$)? Why does this happen?

c. What do you notice as x approaches 8 from the right (from values greater than 8, written $x \to 8^+$)? Why does this happen?

d. Use your work in parts (a) through (c), along with your knowledge of how to predict horizontal asymptotes for rational functions, to sketch the graph of f on the given axes.

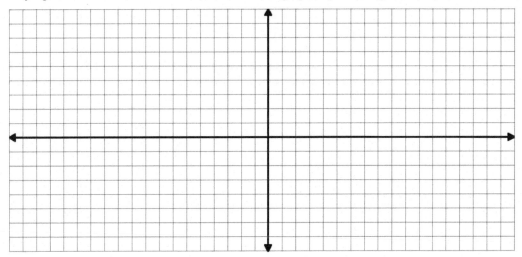

Vertical Asymptotes

A **vertical asymptote** occurs at a real number $x = a$ if

*   as $x \to a^-$ (as x approaches a from the left, or from values less than a), the value of $f(x)$ increases or decreases without bound

and

*   as $x \to a^+$ (as x approaches a from the right, or from values greater than a), the value of $f(x)$ increases or decreases without bound.

---

5. Let $f(x) = \dfrac{p(x)}{r(x)}$ where p and r are polynomial functions.

   a. For what value(s) of x is $f(x)$ undefined?

   b. For what value(s) of x does $f(x) = 0$?

*6. For each of the following functions, predict where any vertical asymptotes will exist. Then test your predictions using a graphing calculator.

   a. $f(x) = \dfrac{8}{x+1}$

   b. $g(x) = \dfrac{x}{(2x-8)^2}$

   c. $h(x) = \dfrac{6x^2 - 24}{2(x-3)(x+1)}$

   d. $m(x) = \dfrac{x+3}{x^2 + 4x - 5}$

   e. $n(x) = \dfrac{x^2 + 1}{x^2 - 1}$

   f. $c(x) = \dfrac{2(x+6)}{x+6}$

7. Five gallons of liquid flavoring are poured into a large vat. Water is added to the vat and mixed with the flavoring to produce a mixture that will be bottled and sold as a drink.

   a. If water is added until the total mixture is 7 gallons, what is the ratio of the volume of flavoring to the volume of water? What does this ratio represent? (Hint: *How many times as large…*)

   7 total − 5 liquid = 2 water

   $\frac{5}{2} = 2.5$ gallons of flavor/gallon of water

   OR 2.5 as much flavor as water

   b. If water is added until the total mixture is 18 gallons, what is the ratio of the volume of flavoring to the volume of water?

   18 − 5 = 13 water

   $\frac{5}{13} = 0.3846154$ gallons flavor/gallon water

   ↓ x as much flavor

c. If x is the **total** volume of the mixture (in gallons), what expression represents the volume of water (in gallons) in the mixture?

$$x - 5$$

d. Define a function f that determines the ratio of flavoring to water in the mixture, R, in terms of the total volume of the mixture (in gallons), x. What values of the domain make sense in this context?

$f(x) = R$ (ratio) in terms of x (total volume of mixture)

$$f(x) = \frac{5}{x-5}$$

$$(5, \infty)$$

e. What does it mean to say, "as $x \to 5^+$" in this context?

As x approaches 5

f. What happens to the value of $f(x)$ as $x \to 5^+$? Why does this make sense?

$f(x)$ approaches ∞. $f(x) \to \infty^+$

less and less water, bigger and bigger ratio

$$\frac{5}{0.0001} = 50,000 \qquad \frac{5}{0.00001} = 500,000, \text{ etc.}$$

g. Graph f using a calculator or graphing software, then complete the following.

i. Complete the following statement: As $x \to \infty$, $f(x) \to$ _____.

ii. Explain why your answer makes sense in this problem context. (*Hint: As the volume of the mixture increases without bound, ...*)

8. Write an explanation describing how to determine the vertical asymptote(s) of a rational function.

9. The graphs of $p(x)$ and $r(x)$ are given. The rational function h is defined by $h(x) = \dfrac{p(x)}{r(x)}$. Let's explore the behavior of h by thinking about the behaviors of p and r.

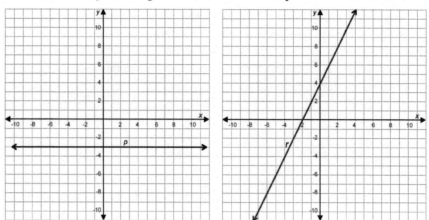

a. As x increases from -10 to -2, describe whether the values of each function are I) positive or negative and II) constant, increasing or decreasing.
   i. The values of $p(x)$ are…   ii. The values of $r(x)$ are…   iii. The values of $h(x)$ are…

b. Evaluate $h(-2)$.

c. As x increases from -2 to 0, describe whether the values of each function are I) positive or negative and II) constant, increasing or decreasing.
   i. The values of $p(x)$ are…   ii. The values of $r(x)$ are…   iii. The values of $h(x)$ are…

d. Evaluate $h(0)$.

e. As x increases from $x = 0$ to $x = 5$, describe the behavior of h.

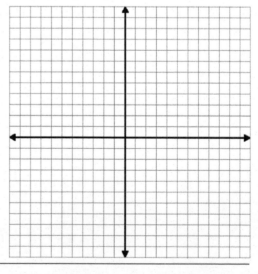

f. Sketch a graph of the function h.

Use the four rational functions defined in (i-iv) to answer Questions 1-4.

i) $f(x) = \dfrac{x-3}{x+2}$ * ii) $g(x) = \dfrac{3x^2}{(x-1)(x-3)}$ * iii) $h(x) = \dfrac{(x+1)(x-1)}{x-2}$ * iv) $k(x) = \dfrac{x-2}{x^2-4}$

*1. a. Find the real zeros (or roots) of each function.

    i)   ii)   iii)   iv)

    b. Describe your method for determining the zeros and what they represent.

*2. a. Determine the vertical intercept of each function.

    i)   ii)   iii)   iv)

    b. Describe your method for determining the vertical intercepts and what they represent.

*3. a. What is the domain of each function?

    i)   ii)   iii)   iv)

    b. For what values of x that are excluded from a function's domain does the graph of the function have a hole instead of a vertical asymptote? Explain how you know that the graph has a hole.

*4. a. Determine the end behavior and horizontal asymptote for each function.

i) $f(x) = \dfrac{x-3}{x+2}$ * ii) $g(x) = \dfrac{3x^2}{(x-1)(x-3)}$ * iii) $h(x) = \dfrac{(x+1)(x-1)}{x-2}$ * iv) $k(x) = \dfrac{x-2}{x^2-4}$

i) As $x \to \infty, f(x) \to$ _____

    As $x \to -\infty, f(x) \to$ _____

    Horizontal Asymptote: _____

ii) As $x \to \infty, g(x) \to$ _____

    As $x \to -\infty, g(x) \to$ _____

    Horizontal Asymptote: _____

iii) As $x \to \infty, h(x) \to$ _____

    As $x \to -\infty, h(x) \to$ _____

    Horizontal Asymptote: _____

iv) As $x \to \infty, k(x) \to$ _____

    As $x \to -\infty, k(x) \to$ _____

    Horizontal Asymptote: _____

b. Use the information from Exercises #1-4 to sketch a graph of each of the four functions and then check your work with a graphing calculator. (*Remember that if you are ever unclear about how a function behaves on a given interval, you can evaluate the function for values of x.*) ***If your sketch was incorrect, think carefully about why your graph was incorrect.***

i)

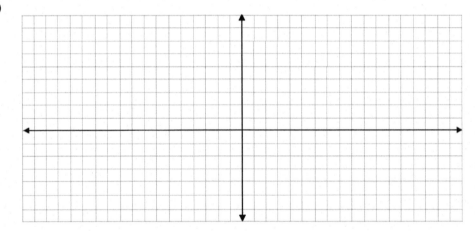

ii)

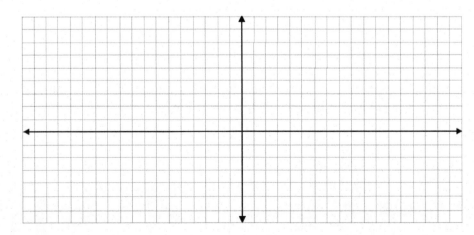

iii)

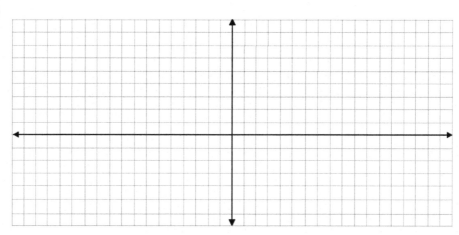

iv)

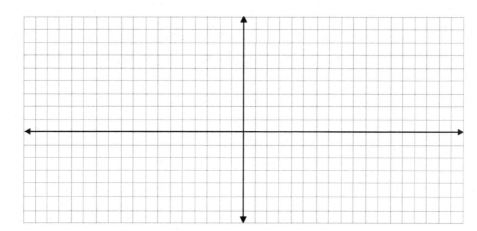

5. Sketch a graph of each of the following rational functions without using a graphing calculator by determining each function's: i) roots; ii) vertical intercept; iii) vertical asymptote(s) (if any); iv) horizontal asymptote(s) (if any); and (v) the sign of the function on intervals of the function's domain.

> *(Remember that if you are ever unclear about the value of a function or how the function behaves on a given interval, you can evaluate the function for values of x in that interval.)*

a. $g(x) = \dfrac{x+7}{x-5}$

b. $h(x) = \dfrac{x^2-9}{x+4}$

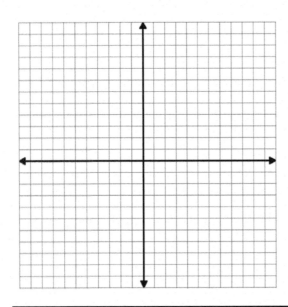
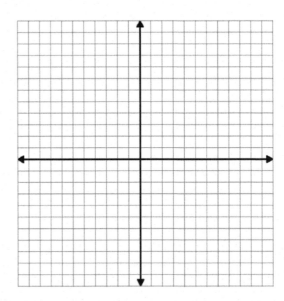

Limit Notation

Limit notation is a concise way to communicate a function's behavior. For a function $y = f(x)$, if $f(x)$ approaches a specific constant value a as x approaches a specific value b (or increases or decreases without bound), we say that a is a **limit** of the function values.

For example, consider the function $f(x) = \frac{x-3}{x+2}$. As x increases without bound ($x \to \infty$), the function value approaches 1. Therefore, we can write "$\lim\limits_{x \to \infty} f(x) = 1$" (read: "As x increases without bound, the function value approaches 1"). In addition, as x decreases without bound, the function value also approaches 1. We can write this as $\lim\limits_{x \to -\infty} f(x) = 1$.

There is a vertical asymptote at $x = -2$, and by doing a bit of work, we can show that $f(x)$ increases without bound as $x \to -2^-$. Since the output value does not approach a specific value, a limit does not exist, or $\lim\limits_{x \to -2^-} f(x)$ DNE. Similarly, $\lim\limits_{x \to -2^+} f(x)$ DNE since, as $x \to -2^+$, $f(x)$ decreases without bound.

*6. The cost of producing coffee mugs to sell at a local coffee shop is $4.75 per mug plus $240 for the manufacturer to construct the mold for making the mugs.

    a.   Define a function *f* that represents the cost (in dollars) of producing *x* mugs.

    b.   Define a function *g* that represents the average cost per coffee mug produced. (Note that determining the *arithmetic mean* or *average cost,* such as the average cost per coffee mug, is different from computing an *average rate of change.*)

    c.   Determine $\lim_{x \to \infty} g(x)$ (if it exists) and explain what this value represents in this context.

    d.   Determine the values of the domain that make sense and explain what this domain represents in this context.

    e.   Determine $\lim_{x \to 0^+} g(x)$ (if it exists) and explain what this limit represents. Is your answer meaningful in this context? Explain.

*7. Use the given functions to determine the value of each limit. If a value does not exist, write DNE.

$$f(x) = \frac{x+3}{x+6} \qquad g(x) = \frac{3x^2}{(x-1)(x-3)} \qquad h(x) = \frac{x^2+1}{x-2} \qquad k(x) = \frac{5x}{x^2-4}$$

  a. $\lim_{x \to \infty} f(x) = $ _____   b. $\lim_{x \to \infty} g(x) = $ _____   c. $\lim_{x \to \infty} h(x) = $ _____   d. $\lim_{x \to -\infty} k(x) = $ _____

      $\lim_{x \to -\infty} f(x) = $ _____ $\lim_{x \to 1^-} g(x) = $ _____ $\lim_{x \to -\infty} h(x) = $ _____ $\lim_{x \to -2^-} k(x) = $ _____

      $\lim_{x \to -6^-} f(x) = $ _____ $\lim_{x \to 1^+} g(x) = $ _____ $\lim_{x \to 2^-} h(x) = $ _____ $\lim_{x \to 2^-} k(x) = $ _____

      $\lim_{x \to -6^+} f(x) = $ _____ $\lim_{x \to 3^+} g(x) = $ _____ $\lim_{x \to 2^+} h(x) = $ _____ $\lim_{x \to 2^+} k(x) = $ _____

8. For each part, you are given information about a rational function. Use the information to sketch a possible graph of the function.

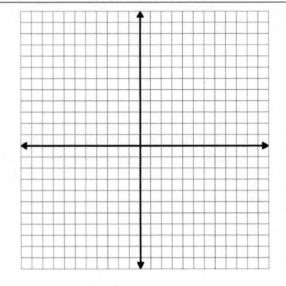

a. $\lim_{x \to \infty} f(x) = -3$

   $\lim_{x \to -\infty} f(x) = -3$

   $\lim_{x \to 2^-} f(x)$ DNE

   As $x \to 2^-$, $f(x)$ increases without bound.

   $\lim_{x \to 2^+} f(x)$ DNE

   As $x \to 2^+$, $f(x)$ decreases without bound.

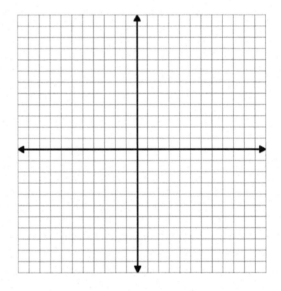

b. $\lim_{x \to \infty} g(x)$ DNE

   As $x \to \infty$, $g(x)$ increases without bound.

   $\lim_{x \to -\infty} g(x)$ DNE

   As $x \to -\infty$, $g(x)$ decreases without bound.

   $\lim_{x \to -4^-} g(x)$ DNE

   As $x \to -4^-$, $g(x)$ decreases without bound.

   $\lim_{x \to -4^+} g(x)$ DNE

   As $x \to -4^+$, $g(x)$ increases without bound.

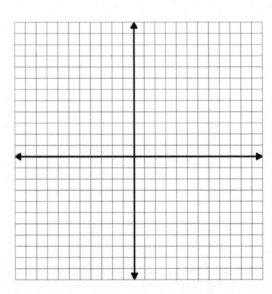

c. $\lim_{x \to \infty} h(x) = 2$ and $\lim_{x \to -\infty} h(x) = 2$

   $\lim_{x \to -3^-} h(x)$ DNE

   As $x \to -3^-$, $h(x)$ increases without bound.

   $\lim_{x \to -3^+} h(x)$ DNE

   As $x \to -3^+$, $h(x)$ decreases without bound.

   $\lim_{x \to 1^-} f(x)$ DNE

   As $x \to 1^-$, $h(x)$ decreases without bound.

   $\lim_{x \to 1^+} f(x)$ DNE

   As $x \to 1^+$, $h(x)$ increases without bound.

*1. Given the graph of the function f,
   determine:
   a. i. the root(s) of f;

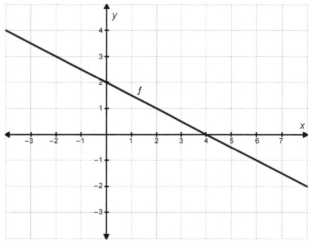

   ii. the y-intercept of f;

   iii. the interval(s) on which f is increasing;   iv. the interval(s) on which f is decreasing;

   v. the interval(s) on which $f(x) < 0$;   vi. the interval(s) on which $f(x) > 0$;

   b. Consider the function g defined by $g(x) = \dfrac{1}{f(x)}$ and think about how x and $g(x)$ change
   together to answer the following.

   i. What does the value of $g(x)$ approach   ii. What does the value of $g(x)$ approach
       as $x \to 4^-$?   as $x \to 4^+$?

   iii. What does the value of $g(x)$ approach   iv. What does the value of $g(x)$ approach
        as $x \to +\infty$ (x increases without bound)?   as $x \to -\infty$ (x decreases without bound)?

   v. Sketch a graph of g on the given axes.

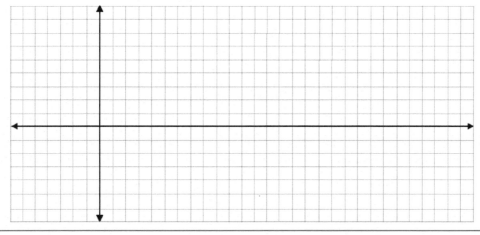

c. How would the graph of g change if g were defined by $g(x) = \dfrac{10}{f(x)}$ instead of $g(x) = \dfrac{1}{f(x)}$?

Explain your reasoning.

d. How would the graph of g change if g were defined by $g(x) = \dfrac{1}{f(x)+1}$ instead of

$g(x) = \dfrac{1}{f(x)}$? Explain your reasoning.

*2. A collection of cylindrical containers (the interior cylinder in the drawing) are being designed to carry different volumes of liquid. Before shipping the container with liquid, it is insulated by packing 2 inches of insulation inside of a cylindrical shipping box and around the sides of the cylindrical container as shown. *Note that a cylinder's volume is represented by the expression $\pi r^2 h$, where r is the radius length of the circular base and h is the cylinder's height.*

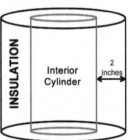

a. If the shipping box's radius length is 4.5 inches long, what is the interior cylinder's radius length? What if the radius of the shipping box is 5.9 inches? x inches?

b. Let x represent the shipping box's radius length (in inches).

   i. Write an expression that represents the interior cylinder's radius length (in inches) in terms of x.

   ii. Write an expression that represents the interior cylinder's volume (in cubic inches) in terms of x and h (the cylinder's height in inches).

c. How tall must the shipping box (and container) be if the shipping box's radius is 4.5 inches long and the interior cylinder holds exactly 200 in^3 of liquid? Repeat if the shipping box's radius is 5.9 inches instead (but the interior cylinder still holds exactly 200 in^3 of liquid).

d. Define a function f that represents the interior cylinder's height (in inches), h, in terms of the shipping box's radius length (in inches), x, assuming the interior cylinder must hold exactly 200 in^3 of liquid.

e. Graph f using a graphing calculator or graphing program, then complete the following.

   i. As $x \to 2^+$, how does the value of $h(x)$ vary? What does this convey about how quantities in the context of this problem vary together?

   ii. As $x \to \infty$, how does the value of $h(x)$ vary?

*3. Given the graph of f, sketch the graph of g defined by $g(x) = \dfrac{5}{f(x)}$.

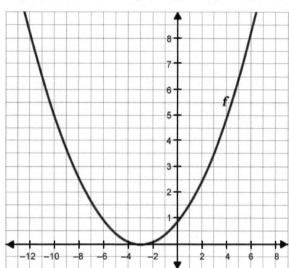
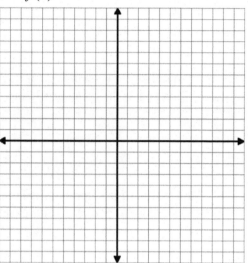

4. Given the graphs of f and g, sketch the graph of h defined by $h(x) = \dfrac{f(x)}{g(x)}$.

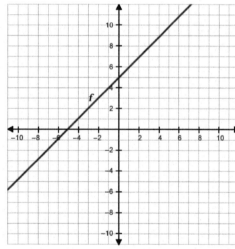
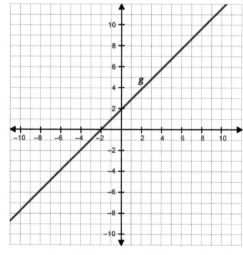

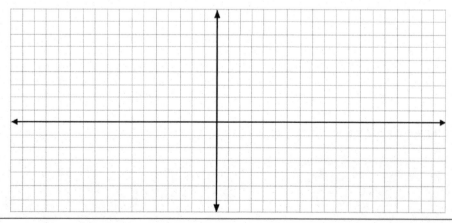

I. RATIONAL FUNCTIONS AND THEIR END BEHAVIOR (Text: S1, 2, 3, 5)

1. The National Center for Education Statistics (nces.ed.gov) keeps careful records of the number of degrees awarded in the United States. The given table shows the number of PhDs awarded to men and women in the U.S. over time.

   - Let $f(t)$ represent the number of PhDs awarded to men in terms of the year, t.
   - Let $g(t)$ represent the number of PhDs awarded to women in terms of the year, t.

   a. What is the value of $f(1940)$? $g(1890)$?
   b. Let h be the function that inputs the year, t, and outputs the value of the ratio $\frac{f(t)}{g(t)}$. That is, $h(t) = \frac{f(t)}{g(t)}$. What is the value of $h(1880)$ and what does it represent in this context?

year, t	# of PhDs awarded to men in the U.S., $f(t)$	# of PhDs awarded to women in the U.S., $g(t)$	Ratio of PhDs awarded to men compared to women, $h(t) = \frac{f(t)}{g(t)}$
1880	51	3	
1890	147	2	
1900	359	23	
1920	522	93	
1940	2,861	429	
1960	8,801	1,028	
1980	69,526	26,105	
2000	64,171	55,414	
2010	76,605	81,953	
2020	85,370	108,690	

   c. As t varies from 1890 to 1900, does the value of $h(t)$ increase or decrease? Why?
   d. Create a graph and plot the points $(1880, h(1880))$, $(1890, h(1890))$, and $(1900, h(1900))$. Label your axes as appropriate. *Note: Use a "broken" horizontal axis to allow the closest tick mark to the vertical axis to represent 1870.*
   e. What must be true in this context if $h(t) = 3$ for some value of t? If $h(t) = 1$?
   f. Is it possible for the output of h to be less than 1 in this context? Explain.
   g. Is it possible for the output of h to be negative in this context? Explain.
   h. Complete the final column in the table to record values of $h(t)$ and then plot each of the entries in the column as points on your graph from part (d). Describe the general behavior of h and what that behavior indicates about this context.

2. The given table shows values of two functions. Function f inputs the year and outputs the average annual compensation (in dollars) for corporate CEOs in the U.S. Function g inputs the year and outputs the average annual compensation (in dollars) for private sector workers. *Note that the values are adjusted for inflation. Data collected from the Economy Policy Institute.*

   a. Evaluate $f(2000)$ and $g(1989)$ and explain what they represent.
   b. The output of function h is the value of the ratio $\frac{f(t)}{g(t)}$. That is, $h(t) = \frac{f(t)}{g(t)}$. Evaluate $h(1978)$ and $h(1995)$ and explain what the values represent in this context.

year, t	average CEO compensation (dollars) $f(t)$	average private sector worker compensation (dollars) $g(t)$	ratio of average annual CEO compensation to average private sector worker compensation $h(t) = \frac{f(t)}{g(t)}$
1965	1,072,000	48,710	
1973	1,406,000	57,380	
1978	1,926,000	58,680	
1989	3,593,000	56,000	
1995	6,968,000	56,010	
2000	25,237,000	59,120	
2007	22,195,000	61,540	
2009	11,924,000	63,950	
2021	29,568,000	68,750	
2022	25,186,000	67,770	

   c. As t increases from 1978 to 1995, did the values of $f(t)$ and $g(t)$ increase, decrease, or remain constant? What does that tell us about this context?

   d. As t increases from 1978 to 1995, did the value of $h(t)$ increase, decrease, or remain constant? What does that tell us about this context?

   e. What would it mean for $h(t) = 10$ for some value of t? What would it mean for $h(t) = 1$ for some value of t?

   f. Is it possible for $h(t)$ to be less than 1 for some value of t?

   g. Use the information in the table to create a table of values for function h. Describe how values of $h(t)$ change as t increases and what that tells you about this context.

3. A beverage company has just completed brewing a large batch of tea (1500 gallons). They will add corn syrup to the tea, then bottle it and prepare it for distribution. The function f relates the ratio R of tea to corn syrup in the beverage when x gallons of corn syrup are added and is defined by $f(x) = \frac{1500}{x}$ with $R = f(x)$.

   a. Complete the given table of values for this function.

   b. What happens to the value of R as x increases without bound and what does this represent in the context?

x	$R = f(x)$
100	
10,000	
1,000,000	
10,000,000	

4. Clark's Soda Company incurred a start-up cost of $82,760 for equipment to produce a new soda flavor. The cost of producing the drink is $0.26 per can. The company sells the soda for $0.75 per can. (*Assume that Clark's Soda Company is able to sell every can produced.*)

   a. Define a function f to determine the cost (including the start-up cost) measured in dollars to produce x cans of soda.

   b. Define a function g to determine the revenue (in dollars) generated from selling x cans of soda.

   c. Define a function h to determine the profit (in dollars) from selling x cans of soda (recall that profit = revenue − cost).

   d. Define a function A to determine the average profit (in dollars per can) of producing x cans of the new soda flavor.

   e. How does $A(x)$ change as the number of cans produced gets larger and larger? Will the average profit per can $A(x)$ ever reach a maximum value? Explain.

5. A salt cell is cleaned using a mixture of water and acid. There are currently 10 liters of water in a bucket. A technician adds varying amounts of acid to the 10 liters of water.

   - Let x = the number of liters of acid added to the 10 liters of water.
   - Let A = the level of acidity of the mixture as a percentage (the percentage of the solution that is acid).

   a. Complete the given table of the mixture's acidity (*measured as a percentage of the total volume*) as acid is added to the 10 liters of water.

   b. Define a function to determine the acidity of the mixture (measured as a percentage) in terms of the number of liters of acid that has been added, x.

   c. Use a calculator to graph f.

liters of acid, x	acidity of mixture, A (as a percentage)
0	
0.5	
1.0	
1.5	
2.0	
2.5	
3.0	
3.5	

   d. Describe how the acidity of the mixture changes as the number of liters of acid added to the solution increases without bound.

   e. Is it possible for the mixture to ever reach 100% acid? Explain.

   f. If the mixture is too acidic, it can damage the salt cell. If the mixture is not acidic enough, it will not clean the cell effectively. It turns out the mixture is most efficient if the acidity is between 40% and 60%. What is the range of liters of acid that should be added to the bucket of water so that the mixture is most efficient?

   g. Define a function that accepts the desired percentage of acid as input and determines the number of liters of acid that should be added to the water as output. (That is, define the formula for f^{-1}.)

6. For each given function, state the horizontal asymptote (if one exists).

   a. $f(x) = \frac{x}{2x-3}$
   b. $g(x) = \frac{x^2}{x+5}$
   c. $h(x) = \frac{4x+1}{2x-10}$
   d. $m(x) = \frac{x^2-2}{x^2+3x+2}$
   e. $n(x) = \frac{17x+200}{10x^3-100x^2}$

   f. $p(x) = \frac{5x^2}{x(x-4)}$
   g. $q(x) = \frac{2x^3+7}{5x^4-10}$
   h. $r(x) = \frac{4x^2+x+11}{(3x+1)(2x-3)}$
   i. $w(x) = \frac{(x+9)(x-5)}{(x+6)(x+2)(x-3)}$

II. RATIONAL FUNCTIONS AND VERTICAL ASYMPTOTES (Text: S1, 2, 4, 6)

7. A weight loss center is so confident in their program that they created a special pricing plan. Joining the program costs a one-time fee of $49.99, and members don't have to pay anything more unless they lose weight. However, members pay the center $2.99 per pound they lose while in the program.

   a. Define a function f to express the cost C for someone joining the program and losing x pounds. What is the practical domain of this function?

   b. We are interested in determining the average cost per pound lost for someone in the program. (*Note that the average cost is NOT an average rate of change.*) Define a function g that represents a member's average cost per pound lost (measured in dollars) as a function of the number of pounds lost. Restrict yourself to consider only values of x such that $x \geq 0$. (*What calculation determines the average cost and why?*)

   c. Complete the given table and then determine the average cost per pound lost for a member losing x pounds while in the program ($x \geq 0$).

x	Cost (dollars) $f(x)$	Average Cost (dollars per pound) $g(x)$
0.01		
0.1		
1		
10		
100		

   d. When x is positive and getting very close to 0 (which we write as $x \to 0^+$), what happens to the average cost per pound? Why?

8. A beverage company has just completed brewing a large batch of tea (1500 gallons). They will add sweetener to the tea, then bottle it and prepare it for distribution.

   a. Suppose 10 gallons of corn syrup is added to the mixture as a sweetener. What is the ratio of tea to corn syrup in the mixture? What if 50 gallons of corn syrup is added instead? 200 gallons?

   b. Define a function f whose input x is the volume of corn syrup added (in gallons) and whose output R is the ratio of the volume of tea to the volume of corn syrup in the mixture.

x	$R = f(x)$
0.001	
0.01	
0.1	
1	

   c. What is the practical domain for the function you defined?

   d. Complete the given table of values for this function.

   e. What happens to R as $x \to 0^+$? Why does this make sense?

9. Given that $f(x) = \frac{x}{x-9}$, answer the questions below.

   a. Complete the given tables of values. Show the calculations that provided your answers.

x	$f(x)$	x	$f(x)$
8		9.001	
8.9		9.01	
8.99		9.1	
8.999		10	

   b. How do the output values of $f(x)$ change as $x \to 9^-$? Why?

   c. How do the output values of $f(x)$ change as $x \to 9^+$? Why?

   d. Using a calculator, graph f. Explain how the graph supports your conclusions above.

   e. What changes if the definition of f becomes $f(x) = -\frac{x}{x-9}$? Why does this happen?

   f. What changes if the definition of f becomes $f(x) = \frac{x}{x+9}$? Why does this happen?

10. Given that $f(x) = \frac{2}{x+7}$, complete the following.

    a. How do the output values of f change as $x \to -7^-$?
    b. How do the output values of f change as $x \to -7^+$?
    c. Using a calculator, graph f. Explain how the graph supports your conclusions in (a) and (b).
    d. What changes if the definition of f becomes $f(x) = -\frac{2}{x+7}$? Why does this happen?

11. Given that $g(x) = \frac{5x}{(x+1)(x-3)}$, complete the following.

    a. How do the output values of g change as $x \to -1^-$? As $x \to -1^+$?
    b. How do the output values of g change as $x \to 3^-$? As $x \to 3^+$?
    c. Using a calculator, graph f. Explain how the graph supports your conclusions in (a) and (b).

12. For each of the following functions, predict where a vertical asymptote will exist and then test your prediction using a graphing calculator.

    a. $f(x) = \frac{x-10}{x+10}$ b. $g(x) = \frac{x^2}{5x-9}$ c. $h(x) = \frac{x}{x^2+2}$ d. $m(x) = \frac{6}{x^2-1}$ e. $n(x) = \frac{5x+5}{x+1}$ f. $p(x) = \frac{6-x}{(x-4)(x-5)}$

III. GRAPHING RATIONAL FUNCTIONS AND UNDERSTANDING LIMITS (Text: S3, 4, 5, 6)

13. Which of the following best describes the behavior of the function f defined by $f(x) = \frac{1}{(x-2)^2}$?

    Provide a rationale for your answer.
    a. As the value of x increases without bound, the value of f decreases without bound.
    b. As the value of x increases without bound, the value of f increases without bound.
    c. As the value of x increases without bound, the value of f approaches 0.
    d. As the value of x approaches 2, the value of f approaches 0.
    e. (a) and (c)

In Exercises #14-22, identify the x-intercept(s), y-intercept, horizontal asymptotes, vertical asymptotes, and the function's domain. (*State DNE in cases when an intercept or asymptote does not exist.*) Use this information to sketch a graph of the function.

14. $a(x) = \frac{3}{x-7}$ 15. $b(x) = \frac{x}{2x+6}$ 16. $d(x) = \frac{9x}{3x+3}$ 17. $f(x) = \frac{-x+3}{2x-1}$ 18. $g(x) = \frac{9x^2-144}{x^2-1}$

19. $h(x) = \frac{14x}{3x-4}$ 20. $k(x) = \frac{x-11}{x^2-5x+6}$ 21. $p(x) = \frac{x(x+2)}{4x+1}$ 22. $q(x) = \frac{x^2+2x-3}{x^2-1}$

For Exercises #23-24, use the graph of f.

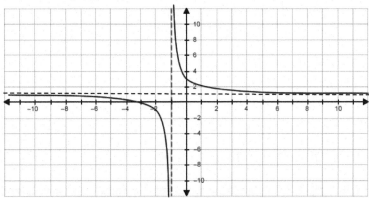

23. Use the graph of f to determine the following limits. If the limit does not exist, write DNE.
    a. $\lim\limits_{x \to \infty} f(x)$ b. $\lim\limits_{x \to -\infty} f(x)$ c. $\lim\limits_{x \to 2^+} f(x)$ d. $\lim\limits_{x \to 2^-} f(x)$

e. $\lim_{x \to 2} f(x)$ f. $\lim_{x \to -1^+} f(x)$ g. $\lim_{x \to -1^-} f(x)$ h. $\lim_{x \to -1} f(x)$

24. Using the graph of f, identify the vertical and horizontal intercepts, the vertical and horizontal asymptotes, and the domain and range of the function. Then, determine a possible rule for the function f.

For Exercises #25-26, use the graph of g.

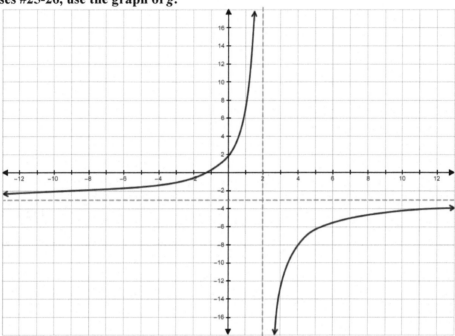

25. Use the graph of g to determine the following limits. If the limit does not exist, write DNE.

     a. $\lim_{x \to \infty} g(x)$ b. $\lim_{x \to -\infty} g(x)$ c. $\lim_{x \to 2^+} g(x)$ d. $\lim_{x \to 2^-} g(x)$

     e. $\lim_{x \to 2} g(x)$ f. $\lim_{x \to 6^+} g(x)$ g. $\lim_{x \to 6^-} g(x)$ h. $\lim_{x \to 6} g(x)$

26. Use the graph of g to identify the vertical and horizontal intercepts, the vertical and horizontal asymptotes, and the domain and range of the function. Then, determine a possible rule for the function g (i.e., represent the function algebraically).

For Exercises #27-28, use the table of values for the rational functions f, g, and h.

x	−10000	−1000	−100	100	1000	10000
$f(x)$	−10001.9997	−1001.9970	−101.9694	97.9706	997.9970	9997.9997
$g(x)$	−6.0004	−6.0040	−6.0404	−5.9604	−5.9960	−5.9996
$h(x)$	−0.0005	−0.0050	−0.0506	0.0496	0.0050	0.0005

27. Based on the table of values, explain the end behavior of each function.

     a. $\lim_{x \to \infty} f(x)$ b. $\lim_{x \to -\infty} f(x)$ c. $\lim_{x \to \infty} g(x)$ d. $\lim_{x \to -\infty} g(x)$ e. $\lim_{x \to \infty} h(x)$ f. $\lim_{x \to \infty} h(x)$

28. Consider the rules that define the functions f, g, and h. How must the degree of the numerator compare to the degree of the denominator for each function (e.g., greater than, less than, or equal to)?

For Exercises #29-30, use the graph of *h*.

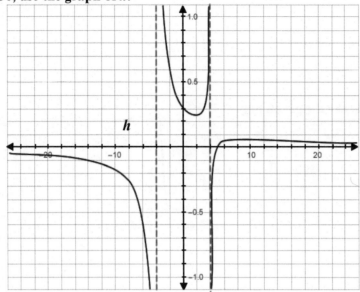

29. Use the graph of *h* to determine the following limits. If the limit does not exist, write DNE.

    a. $\lim\limits_{x \to \infty} h(x)$

    b. $\lim\limits_{x \to -\infty} h(x)$

    c. $\lim\limits_{x \to -4^+} h(x)$

    d. $\lim\limits_{x \to -4^-} h(x)$

    e. $\lim\limits_{x \to -4} h(x)$

    f. $\lim\limits_{x \to -4^+} h(x)$

    g. $\lim\limits_{x \to 4^-} h(x)$

    h. $\lim\limits_{x \to 4} h(x)$

30. Use the graph of *h* to identify the vertical and horizontal intercepts, the vertical and horizontal asymptotes, and the domain and range of the function. Then, determine a possible rule for the function *h*.

31. Use the given information about function *f* to complete parts (a) through (d).

    • $\lim\limits_{x \to \infty} f(x) = 0$ and $\lim\limits_{x \to -\infty} f(x) = 0$.

    • $\lim\limits_{x \to -6^-} f(x)$ DNE because $f(x)$ decreases without bound as $x \to -6^-$.

    • $\lim\limits_{x \to -6^+} f(x)$ DNE because $f(x)$ increases without bound as $x \to -6^+$.

    • $\lim\limits_{x \to 6^-} f(x)$ DNE because $f(x)$ decreases without bound as $x \to 6^-$.

    • $\lim\limits_{x \to 6^+} f(x)$ DNE because $f(x)$ increases without bound as $x \to 6^+$.

    • The *y*-intercept is $(0, \frac{5}{36})$.

    • The *x*-intercept is $(5, 0)$.

    a. Identify the vertical and horizontal asymptotes for *f*. State DNE if none exist.
    b. What is the domain of *f*?   c. Find a possible formula for *f*.   d. Sketch a graph for *f*.

32. Use the given information about function *g* to complete parts (a) through (d).

    • $\lim\limits_{x \to \infty} g(x) = 7$ and $\lim\limits_{x \to -\infty} g(x) = 7$.

    • $\lim\limits_{x \to -2^+} g(x)$ DNE because $g(x)$ decreases without bound as $x \to -2^+$.

    • $\lim\limits_{x \to -2^-} g(x)$ DNE because $g(x)$ increases without bound as $x \to -2^-$.

    • The *y*-intercept is $(0, \frac{7}{2})$.

    • The *x*-intercept is $(-1, 0)$.

    a. Identify the vertical and horizontal asymptotes for *g*. State DNE if none exist.
    b. What is the domain of *g*?   c. Find a possible formula for *g*.   d. Sketch a graph for *g*.

33. Use the given information about function h to complete parts (a) through (d).

  - $\lim\limits_{x \to \infty} h(x)$ DNE because $h(x)$ increases without bound as $x \to \infty$.

  - $\lim\limits_{x \to -\infty} h(x)$ DNE because $h(x)$ decreases without bound as $x \to -\infty$.

  - $\lim\limits_{x \to -(9/2)^+} h(x)$ DNE because $h(x)$ increases without bound as $x \to -(9/2)^+$.

  - $\lim\limits_{x \to -(9/2)^-} h(x)$ DNE because $h(x)$ decreases without bound as $x \to -(9/2)^-$.

  - The y-intercept is $(0, -\frac{10}{9})$.

  - The x-intercepts are $(-\sqrt{5}, 0)$ and $(\sqrt{5}, 0)$.

  a. Identify the vertical and horizontal asymptotes for h. State DNE if none exist.
  b. What is the domain of h?   c. Find a possible formula for h.   c. Sketch a graph for h.

IV. CO-VARIATION OF NUMERATORS AND DENOMINATORS IN RATIONAL FUNCTIONS

34. Use the graph of f to answer the questions that follow.
  a. Describe the important characteristics of f.
  b. Define a new function g as $g(x) = \frac{1}{f(x)}$.

    i.   What happens to $g(x)$ as $x \to 2^-$? As $x \to 2^+$?
    ii. What happens to $g(x)$ as x increases without bound? As x decreases without bound?
    iii. Sketch a graph of g.
  c. What, if anything, changes if we update the definition of g to $g(x) = \frac{10}{f(x)}$? Why does this happen?

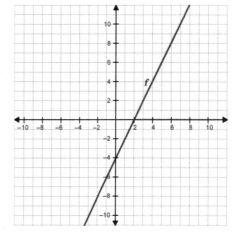

35. Use the graph of f to answer the questions that follow.
  a. Describe the important characteristics of f.
  b. Define a new function g as $g(x) = \frac{2}{f(x)}$.

    i.   What happens to $g(x)$ as $x \to -18^-$? As $x \to -18^+$?
    ii. What happens to $g(x)$ as x increases without bound? As x decreases without bound?
    iii. Sketch a graph of g.
  c. What, if anything, changes if we update the definition of g to $g(x) = -\frac{2}{f(x)}$? Why?

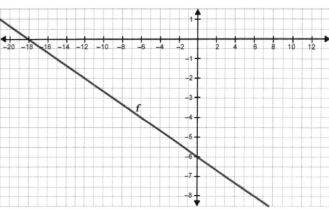

36. Given the graph of *f*, sketch the graph of *g* if $g(x) = \frac{1}{f(x)}$.

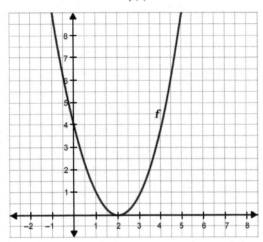

37. Given the graph of *f*, sketch the graph of *g* if $g(x) = \frac{x}{f(x)}$.

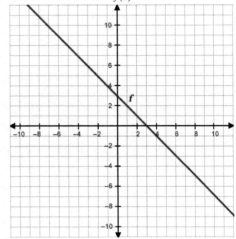

38. Given the graphs of *f* and *g*, sketch the graph of *h* if $h(x) = \frac{g(x)}{f(x)}$.

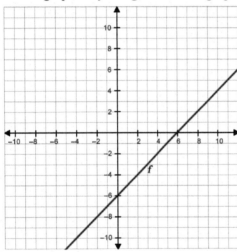
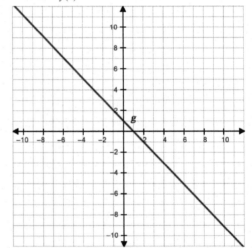

39. Given the graphs of *f* and *g*, sketch the graph of *h* if $h(x) = \frac{f(x)}{g(x)}$.

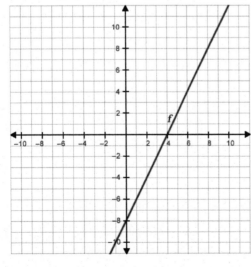
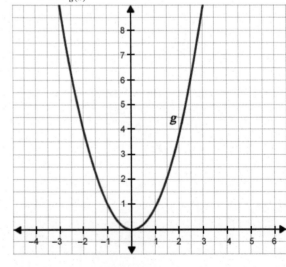

Part 1: The Meaning of Exponents

(Recommended use: Prior to Investigation 4)

1. What does an exponent tell us? For example, what does the expression 3^4 represent?

2. Rewrite each of the following using exponents.
   a. $6 \cdot 6 \cdot 6 \cdot 6 \cdot 6 \cdot 6 \cdot 6$
   b. $x \cdot x \cdot x \cdot x$
   c. $5 \cdot 5 \cdot 5 \cdot 8 \cdot 8 \cdot 8 \cdot 8 \cdot 8$

3. Rewrite each of the following as a product of factors (without using exponents). *Do not simplify your answer.*
   a. $10^2 \cdot 9^3$
   b. $4p^5 t^4$

   c. $\left(5^2\right)^4$
   d. $\left(7x^3 y^2\right)\left(2x^2 y^4\right)$

4. Simplify your results in Exercises #3c and #3d as much as possible.

Part 2: Radicals

(Recommended use: Prior to Investigation 4)

5. What does the square root of a number represent? For example, what is the value of $\sqrt{64}$ and what does it represent?

6. a. Why are there two solutions to the equation $x^2 = 25$?
   b. What are the solutions to the equation $x^2 = 81$? What about $w^2 = 144$?

7. a. Are there also two solutions to the equation $x^3 = 8$? Explain.
   b. What are the solutions to the equation $r^3 = 27$? What about $p^3 = -125$?

8. The *cube root* of a number $\left(\sqrt[3]{\#}\right)$ represents the number that, when raised to an exponent of 3, returns the original number. For example, $\sqrt[3]{8} = 2$ because $2^3 = 8$. Find the value of each of the following.
   a. $\sqrt[3]{64}$
   b. $\sqrt[3]{729}$
   c. $\sqrt[3]{-343}$

9. For some number x, what does each of the following represent?
   a. $\sqrt[5]{x}$
   b. $\sqrt[4]{x}$
   c. $\sqrt[10]{x}$

10. Solve each of the following equations.

a. $2b^2 = 200$ b. $r^4 = 1,296$ c. $x^7 = -128$

d. $5x^3 = 320$ e. $64z^8 = 0.25$ f. $12x^5 + 4 = 2,920$

Part 3: Percent Change and Factors
(Recommended use: Prior to Investigation 5)

11. Suppose an item has an original price of $65 and we purchase it on sale for 20% off.

a. What number can we multiply $65 by to find out the discount in dollars? What is the discount in dollars?

b. Use the result of part (a) to determine the price we are paying.

c. What percent of the original price are we paying?

d. What number can we multiply $65 by to determine the sale price of the item?

12. For each sale described, do the following.
    i) State the number we can multiply the original price by to find the sale price.
    ii) Find the sale price in dollars.

a. original price: $150, sale: 40% off   b. original price: $19, sale: 10% off

c. original price: $915.99, sale: 15% off   d. original price: $22.99, sale: 12.5% off

13. Suppose a store is raising the price of a $42 item by 15%.

a. By how many dollars is the price increasing? What number can we multiply $42 by to determine this?

b. What is the new price of the item?

c. What number can we multiply $42 by to find the new price of the item?

14. For each price increase described, do the following.
    i) State the number we can multiply the original price by to find the new price.
    ii) Find the new price in dollars.

a. original price: $52, increase: 10%   b. original price: $13, increase: 50%

c. original price: $14.99, increase: 2.3%   d. original price: $1,499.99, increase: 100%

15. A city with a population of 480,560 people at the end of the year 2000 grew by 6.2% over 10 years. What was its population at the end of 2010?

16. Student Council reported that attendance at Prom this year was 4.8% less than attendance last year. If 818 people attended Prom last year, how many people attended Prom this year?

A *sequence* in mathematics is an ordered set of objects (usually numbers), such as 2, 4, 6, 8, 10 or 7, 5, 6, 4, 5, 3. The objects in the sequence are called *terms*.

Finite vs. Infinite Sequences

A *finite sequence* is a sequence with a set number of terms. Finite sequences are written like 2, 4, 6, 8, 10 or 2, 4, 6, …, 32.

An *infinite sequence* is a sequence with infinitely many terms where the pattern generating the sequence is repeated without end. Infinite sequences are written like 5, 10, 15, …

---

1. Consider the sequence 1, 2, 4, 8, 16, …
   a. Describe the pattern.

   b. Find the next three terms.

   c. Is it difficult to find the value of the 1000th term? If so, explain why. If not, find its value.

2. Consider the sequence 3, 6, 11, 18, 27, …
   a. Describe the pattern.

   b. Find the next three terms.

   c. Is it difficult to find the value of the $2,500^{th}$ term? If so, explain why. If not, find its value.

3. Consider the sequence $\frac{2}{1}, \frac{5}{4}, \frac{10}{9}, \frac{17}{16}, \frac{26}{25}, …$
   a. Describe the pattern.

   b. Will any of the following be terms in this sequence? If so, which one(s)?
      $\frac{101}{100}$ $\frac{116}{115}$ $\frac{161}{160}$ $\frac{226}{225}$

   c. Is it difficult to find the value of the 50th term? If so, explain why. If not, find its value.

4. Sequences can be thought of as functions in which the input quantity is the ***term position*** [consisting of natural numbers (positive integers)] and the output quantity is the ***term value*** [consisting of the terms of the sequence]. Let's take another look at the sequence in Exercise #3.

input: term position	output: term value
1	2/1
2	5/4
3	10/9
4	17/16
5	26/25

Like other functions, we can create the graph of a sequence. By convention we track the term position on the horizontal axis and the term value on the vertical axis.

a. Before graphing, consider the following question. When you plot the sequence, should the points be connected? Explain.

b. Graph the sequence. Use at least the first six terms.

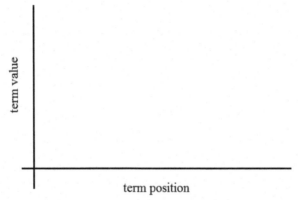

c. What does the graph suggest about the term values as the term position increases?

The Limit of a Sequence

The behavior you noted in Exercise #4 means that this sequence has a ***limit***, or a constant value that the term values approach as the term position increases.[1]

Note: Only infinite sequences (sequences with infinitely many terms) are said to have limits. Finite sequences (sequences with a definitive end) do not have a limit.[2]

In Exercise #4, the *limit is 1*.

---

[1] The formal definition (paraphrased) requires that, for a limit to exist, the term values must be able to get as close as we want to the limiting value as the term position increases for the limit to exist. In other words, if we say that a sequence *has a limit of 4*, it means that at some point the term values must be within (for example) 0.1 of 4, and within 0.003 of 4, and within 0.00000007 of 4, etc. for all remaining terms.

[2] Even if the term values of a finite sequence are getting closer to a constant value, eventually the sequence stops and the last term's value is the closest we can get to this constant. Therefore, the term values can't get arbitrarily close to the constant value. For example, consider the sequence 7.9, 7.99, 7.999, 7.9999. The term values are certainly getting closer to 8, but the final term is 0.0001 away from 8. We could never get within 0.000001 of 8, for example, or within 0.00000003 of 8. Therefore "8" doesn't fit the requirement to be a limit of the sequence.

d. We said that sequences can be thought of as functions. Explain to a partner your understanding of this statement. Then explain how they are different from some of the other functions we have studied.

5. A new sequence is generated by taking the difference of 3 and the term values from the sequence in Exercise #4. For example, the first term of the new sequence is $3 - \frac{2}{1} = 1$. The second term of the sequence is $3 - \frac{5}{4} = \frac{7}{4}$.

a. Graph this new sequence (use at least 6 points).

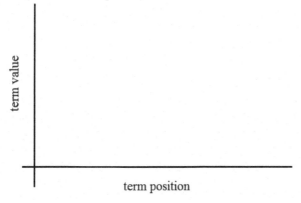

b. Does the sequence appear to have a limit? If so, what is the limit?

c. What is the relationship between the limit of this sequence and the limit of the sequence in Exercise #4? (*That is, is there a mathematical reason why the sequence has term values approaching this limit while the term values of the other sequence approach 1?*)

For Exercises #6-9, do the following.
   a) Examine the pattern and then write the next three terms.
   b) Graph the sequence (use at least six points).
   c) Does the sequence appear to have a limit? If it does, state the limit.

6. 7, 9, 11, 13, ...

7. $1, \frac{1}{2}, \frac{1}{3}, \frac{1}{4}, ...$

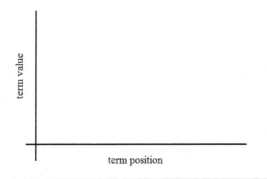

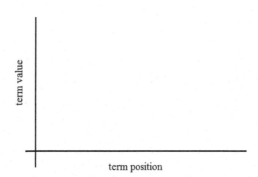

8.   5.1, 4.9, 5.01, 4.99, 5.001, 4.999, ...   9.   1, –1, 2, – 2, 4, – 4, 8, – 8, ...

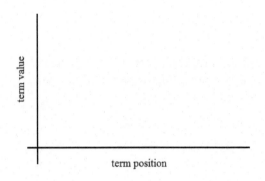

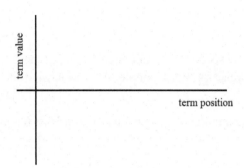

10. A sequence is formed by choosing any real number *x* to be the first term, then generating the
     sequence by taking each term value and dividing by 10 to get the next term value.
     a. Choose a few possible values of *x* and explore the kinds of sequences produced by following this
         pattern.

     b. Do all sequences formed in this manner have a limit? If so, what is the limit? Why does this
         happen?

     c. What if the pattern was instead formed by multiplying the term value by 10 to get the subsequent
         term value? Would sequences formed in this manner have a limit? Explain.

11. Explain, in your own words, what it means for a sequence to have a limit.

In Exercises #1-2, find the indicated term value for the given sequence.

1. $15, 17, 19, 21, 23, 25, 27, \ldots, 73$; find a_6

2. $44, 22, 11, \ldots$; find a_5

3. If a_n represents the value of the nth term in some sequence, how can we represent each of the following?

   a. the value of the term before the nth term

   b. the value of the term after the nth term

   c. the value of the term two terms before the nth term

We've seen that a sequence can be thought of as a function where each term position maps to a single term value. Therefore, it's not surprising that we can also write formulas that relate these quantities.

Recursive Formula for a Sequence's Term Values

4. Consider the sequence $17, 23, 29, 35, \ldots, 323$. One of the most common ways to describe this sequence is by saying something similar to "The sequence begins with 17, and the value of every term is 6 more than the value of the previous term."

   a. Using sequence notation, how can you communicate to someone that the value of the first term in the sequence is 17?

   b. Using sequence notation, how can you communicate to someone that the value of any term (such as the nth term) is always 6 more than the value of the previous term?

Recursive Formula for a Sequence's Term Values

A ***recursive formula*** defines the value of a term a_n based on the value of the previous term or terms.

In Exercises #5-7, use the given formula to write the first four terms of each sequence.

5. $a_1 = 3$
$a_n = a_{n-1} - 5$

6. $a_1 = 40$
$a_n = \dfrac{a_{n-1}}{4}$

7. $a_1 = 1$
$a_2 = 2$
$a_n = (a_{n-1})(a_{n-2})$

In Exercises #8-9, write a recursive formula to define each sequence, then find a_9.

8. $11, 8, 5, 2, \ldots$

9. $-7, 21, -63, 189, \ldots$

10. John wasn't feeling well, so he went to the doctor yesterday. The doctor gave him a prescription for antibiotics and told him to take one 450 mg dose every 8 hours. John's body metabolizes the drug such that 33% of the medicine remains in his body by the time he takes the next dose.
    a. Why might doctors create a schedule where you take a new dose before the previous dose is completely removed from the body?

    b. Write a recursive formula that will tell you how much medicine is in John's body after taking n doses of the medicine.

    c. Suppose John never stops taking the medicine.
       i. What will happen to the amount of medicine in his body after each dose? Perform some calculations if necessary to explore this question.

       ii. Is there a limit to the maximum amount of antibiotics in John's system? If so, what is the limit?

11. What's the biggest drawback of using a recursive formula to describe a sequence?

12. Consider the following sequence defined recursively. Explain to a partner how determining the values in the sequence are like evaluating function composition expressions such as $f(f(4))$.
$$a_1 = 7$$
$$a_n = 3 \cdot a_{n-1} - 10$$

13. Consider the sequence $4, 8, 12, 16, \ldots$. Each term value is exactly 4 times as large as the corresponding term position, so we could think of the sequence as $4(1), 4(2), 4(3), 4(4), \ldots$ and say that $a_n = 4n$. Find the value of the 25th term.

14. Consider the sequence $1, 8, 27, 64, \ldots$. Each term value is the third power of the corresponding term position, so we could think of the sequence as $1^3, 2^3, 3^3, 4^3, \ldots$ and say that $a_n = n^3$. Find the value of the 25th term.

15. How are the formulas $a_n = 4n$ and $a_n = n^3$ different from the recursive formulas we wrote earlier in this investigation?

Explicit Formula for a Sequence's Term Values

An *explicit formula* defines the value of a term a_n based on its position n.

Explicit Formula for a Sequence's Term Values

In Exercises #16-18, find the value of the 12th term in each sequence.

16. $a_n = \dfrac{2}{n}$ 17. $b_n = 0.5(n-1)$ 18. $c_n = \dfrac{3n}{n+1}$

In Exercises #19-20, write the explicit formula defining the sequence.

19. $12, 24, 36, \ldots, 192$ 20. $0, 3, 8, 15, 24, \ldots, 288$

21. How many terms are in the sequences in Exercises #19 and #20?

22. The explicit formulas you wrote Exercises #19 and #20 represent the term value based on its position. Write the inverses for each relationship. (*That is, write the formulas the represent the term's position based on its value.*)

23. When a certain ball is dropped, it bounces back up to ¾ of the distance it fell. Suppose the ball is initially dropped from a height of 12 meters.
    a. Write the first three terms of the sequence describing the height the ball will return to after n bounces.

    b. Write an explicit formula for the sequence that will tell you the height the ball will bounce up to after n bounces.

    c. Is there a limit to this sequence?

Use the following four sequences for Exercises #1-4.
    i) −6, 2, 10, 18, ...   ii) 13, 11, 9, 7, ...   iii) 3.74, 3.79, 3.84, ...   iv) 4.1, 3.3, 2.5, ...

1. Write a recursive formula defining the term values for each sequence.

2. How are the sequences similar to each other?

3. a. How are each of the sequences similar to linear functions?

    b. How are the sequences different from linear functions?

4. Which of the following sequences are similar to the four given sequences? Which are different?
       8, 1, −6, −13, −20, …

       1, 4, 9, 16, 25, …

       3, 6, 12, 24, 48, …

       1, 1.1, 1.2, 1.3, 1.4, …

Arithmetic Sequences

Common Difference:

Arithmetic Sequence:

The Recursive Formula for an Arithmetic Sequence:

Let's take a moment to review formulas for linear functions before thinking about how to write an explicit formula for any arithmetic sequence.

5. Write the formulas for the following linear function relationships.
    a. The constant rate of change of y with respect to x is -2.4 and $(x, y) = (3, 9)$ is one ordered pair for the relationship.

    b. The constant rate of change of y with respect to x is 3 and $(x, y) = (1, 7)$ is one ordered pair for the relationship.

6. Based on her answers to Exercises #3-5, Shelly looked at the following arithmetic sequence and had an idea.

$$7, 10, 13, 16, 19, 22, \ldots$$

"This sequence is kind of like a linear function with a constant rate of change of 3. I can write $\Delta a_n = 3 \cdot \Delta n$ to think about how this sequence works."

    a. What do you think Shelly understands when she writes $\Delta a_n = 3 \cdot \Delta n$?

    b. We know that the first term in the sequence is 7. We could think about this like the ordered pair $(n, a_n) = (1, 7)$. Based on this idea, answer the following questions.
        i. What is the change in the term position from the 1st term to the 12th term?

ii. What is the change in term position from the 1ˢᵗ term to the 24ᵗʰ term?

iii. What is the change in term position from the 1ˢᵗ term to the nᵗʰ term?

c. For any change in the term position away from $n = 1$, by how much does the term value change? How you can represent this idea? [For example, when the term position changes from the 1ˢᵗ term to the 12ᵗʰ term, by how much does the term value change? What about from the 1ˢᵗ term to the nᵗʰ term?]

d. Knowing that the first term of the sequence is 7 and the common difference is 3, find each of the following term values.
   i. The value of the 54ᵗʰ term.

   ii. The value of the 380ᵗʰ term.

   iii. The value of the nᵗʰ term for any value of n.

7. Write an explicit formula for the term values in each of the following arithmetic sequences.
   a. 19, 15, 11, 7, …

   b. −56, −41, −26, −11, …

   c. The first term is 5.6 and the common difference is 1.3.

d. The first term is a_1 and the common difference is d.

e. The fifteenth term is 84 and the common difference is 5.

The Explicit Formula for an Arithmetic Sequence

For an arithmetic sequence with a common difference of d, the explicit formula for the term values is…

---

8. a. Each of the following are portions of arithmetic sequences. Fill in the blanks and then write an explicit formula and a recursive formula representing the term values in each sequence.
   i. 15, ____, 20, …   ii. 49, –7, ____, , …

   b. Find the 6th term in each sequence.

   c. For the sequence in part (a) given as 15, ____, 20, …, one of the terms in the sequence is 132.5. What is this term's position?

9. a. Each of the following is a portion of an arithmetic sequence. Fill in the blanks and then write an explicit formula and a recursive formula representing the term values in each sequence.

      i.   ____, 32, 40, …   ii. 3, ____, ____, ____, −6, …

    b. Find the 12th term in each sequence.

    c. For the sequence in part (a) given as ____, 32, 40, …, one of the terms in the sequence is 3,472. What is this term's position?

10. Auditoriums are often designed so that there are fewer seats per row for rows closer to the stage. Suppose you are sitting in Row 22 at an auditorium and notice that there are 68 seats in your row. It appears that the row in front of you has 66 seats and the row behind you has 70 seats. Assume this pattern continues throughout the auditorium.

    a. How many seats are in Row 1?

    b. The last row has 100 seats. How many rows are in the auditorium?

    c. Write a recursive formula that defines the value of a_n, the number of seats in row n.

    d. Write an explicit formula that defines the value of a_n, the number of seats in row n.

11. The explicit formula for the term values of a certain sequence is $a_n = 6(n-1) - 27$.

   a. The calculations to determine the 50th term value are given. Answer the questions that go with these calculations.

   $a_n = 6(n-1) - 27$

   $a_{50} = 6(50-1) - 27$ • Step 1

   $a_{50} = 6(49) - 27$ • Step 2

   $a_{50} = 294 - 27$ • Step 3

   $a_{50} = 267$ • Step 4

   i. In Step 1, what does the expression $50 - 1$ represent?

   ii. In Step 2, what does the expression $6(49)$ represent?

   b. [*Inverses*] Solve the formula $a_n = 6(n-1) - 27$ for n and explain what questions this formula helps you to answer.

12. [*Systems of Equations*] The explicit formulas for the term values of two different sequences are $a_n = 4(n-1) + 18$ and $b_n = 6(n-1) - 70$. Is there a term in one sequence that shares the same value and term position as a term in the other sequence? If so, give their value and position.

13. An arithmetic sequence has terms $a_6 = 85$ and $a_{31} = 10$. Write the explicit formula defining the value of a_n.

[Investigation 0 contains review/practice with exponents, the meaning of radical expressions, and how to solve basic equations involving exponents. You can review these concepts as needed.]

Use the following four sequences for Exercises #1-4.

i) $4, 12, 36, ...$ ii) $-\frac{1}{2}, \frac{3}{2}, -\frac{9}{2}, ...$ iii) $6, 2, \frac{2}{3}, ...$ iv) $\frac{10}{3}, -\frac{20}{9}, \frac{40}{27}, ...$

1. Write a recursive formula defining the term values for each sequence.

2. How are the sequences similar to one another?

Geometric Sequences

Common Ratio:

Geometric Sequence:
Recursive Formula for a Geometric Sequence:

---

4. Consider the following sequence.

$$3, 6, 12, 24, 48, \underline{\hspace{1.5cm}}, \underline{\hspace{1.5cm}}, \underline{\hspace{1.5cm}}, ...$$

a. Verify that this is a geometric sequence and fill in the next three terms.

b. How does the term value change when the term position changes? Let's explore.
   i. When the term position changes by 1, how does the term value change?

   ii. When the term position changes by 2, how does the term value change?

---

iii. When the term position changes by 3, how does the term value change?

iv. When the term position changes by 7, how does the term value change?

v. When the term position changes by –1, how does the term value change?

c. What is the change in term position from the 1st term to the 40th term? Use your answer to write an expression representing the value of the 40th term in this sequence.

d. What is the change in term position from the 1st term to the 315th term? Use your answer to write an expression representing the value of the 315th term in this sequence.

e. What is the change in term position from the 1st term to the nth term? Use your answer to write an expression representing the value of the nth term in this sequence.

5. Consider the following sequence.

$$4, -12, 36, -108, 324, \underline{\hspace{2cm}}, \underline{\hspace{2cm}}, \underline{\hspace{2cm}}, \ldots$$

a. Verify that this is a geometric sequence and fill in the next three terms in the sequence.

b. How does the term value change when the term position changes? Let's explore.
   i. When the term position changes by 1, how does the term value change?

   ii. When the term position changes by 4, how does the term value change?

   iii. When the term position changes by –2, how does the term value change?

   iv. When the term position changes by –3, how does the term value change?

c. What is the change in term position from the 1st term to the 24th term? Use your answer to write an expression representing the value of the 24th term in this sequence.

d. What is the change in term position from the 1st term to the 217th term? Use your answer to write an expression representing the value of the 217th term in this sequence.

e. What is the change in term position from the 1st term to the nth term? Use your answer to write an expression representing the value of the nth term in this sequence.

6. Suppose $450 is invested in an account earning 2.8% interest compounded annually. Furthermore, suppose that no additional deposits or withdrawals are made.

a. To find the account value after the first year we multiply $450 by 1.028. What is the value of the account after 1 year?

b. If we make a list of the account balance at the end of each year since the initial deposit was made, why will this list form a geometric sequence?

c. Write a recursive formula for a_n, the value of the account n years since the initial deposit was made.

d. Write an explicit formula for a_n.

7. If $a_1, a_2, a_3, ...$ is a geometric sequence with a common ratio r, what is the explicit formula for the term values in the sequence?

The Explicit Formula for a Geometric Sequence

For a geometric sequence with a common ratio of r, the explicit formula for the term values is...

---

8. Explain what each of the following represents in the explicit formula (be clear and specific).
   a. n

   b. a_n

   c. $n-1$

   d. r^{n-1}

   e. $a_1 \cdot r^{n-1}$

In Exercises #9-12, do the following.
   a) Verify that the series is geometric.
   b) Write a recursive formula for the term values of the sequence and then write an explicit formula for the term values of the sequence.

9. 5, 20, 80, 320, ...

10. 1458, 486, 162, 54, ...

11. $-8, 12, -18, 27, \ldots$

12. 15, 6, 2.4, 0.96, ...

13. a. Each of the following is a portion of a geometric sequence. Fill in the blanks and then write an explicit formula for each sequence.
   i. 2, 8, ____, ...

   ii. 81, ___, 9, ...

   b. Find the value of the 6th term of each sequence.

14. a. Each of the following is a portion of a geometric sequence. Fill in the blanks and then write an explicit formula for each sequence.

   i. ____, 60, 15, ...

   ii. 2, ____, ____, 250, ...

   b. Find the value of the 12th term of each sequence.

15. A pattern is formed according to the following instructions. Beginning with an equilateral triangle the midpoints of the sides are connected forming four smaller equilateral triangles. The middle triangle is then colored black to create the diagram in Stage 2. This process is repeated with all of the white triangles at each stage to form the diagram in the next stage.

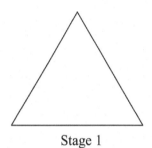

Stage 1

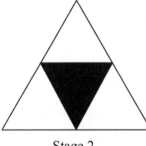

Stage 2

Stage 3

Stage 4   ...

   a. Create a sequence that shows the total number of white triangles a_n at stage n and then write the explicit formula for a_n.

   b. Does the sequence from part (a) have a limit? What does this represent in the context?

c. Assume that the original white triangle has an area of 8 cm². Write the first several terms b_n of the sequence representing the area of one white triangle at stage n and then write the explicit formula for b_n.

d. Still assuming that the original white triangle has an area of 8 cm², write the first several terms c_n of the sequence representing the total area of all of the white triangles at stage n and then write the explicit formula for c_n.

e. Does the sequence from part (d) have a limit? What represent in the context?

f. The term values of a new sequence are defined by $d_n = 8 - c_n$.
   i. What does this new sequence represent?

   ii. Does this sequence have a limit? What does this represent in the context?

16. A geometric sequence has terms $a_9 = 45,927$ and $a_{15} = 33,480,783$. Write the explicit formula defining the value of a_n.

1. After graduating from college, Andy received job offers from two firms. Firm A offered Andy an initial salary of $52,000 for the first year with a 6% pay raise guaranteed for the first five years. Firm B offered Andy an initial salary of $55,000 with a 4% pay raise guaranteed for the first five years. Andy plans to work for five years and then quit and go to graduate school. He really likes both firms, and the hours and job responsibilities are similar, so he will make his decision based on salary.
   a. *Make a prediction:* Which offer do you think he should accept?

   b. What number do we multiply $52,000 by to find out Andy's salary during his second year if he works for Firm A? What number do we multiply $55,000 by to find out Andy's salary during his second year for Firm B?

   c. Write out the sequences that represent his annual salaries with each firm for the first five years.

Year n	1	2	3	4	5
Salary with Firm A in year n					
Salary with Firm B in year n					

   d. Compare Andy's salary in the fifth year working for each firm. Does this tell him which firm he should choose? Why or why not?

   e. Fill in the following table that keeps track of Andy's total salary after working for 1, 2, 3, 4, and 5 years with each firm.

year	total salary earned while working for Firm A by the end of the year	total salary earned while working for Firm B by the end of the year
1		
2		
3		
4		
5		

   f. Pick one row from the table and explain what it represents.

   g. According to Andy's criteria, which firm should he choose?

Series

Series:

_____

Ex: Given the sequence 3, 6, 9, 12, 15, 18, the corresponding series is

---

2. Write the corresponding series for each given sequence and give the total sum.
   a. 1, 6, 3, 14, 10, 29   b. $-7, -2, 2, 5, 7, 8, 8$

When working with series we are often curious about sum of the terms up to a certain point. We call these the *partial sums* and denote them in the form S_n, where S_n is the sum of the first n terms of the sequence.

Partial Sum of a Series

The nth Partial Sum: [represented by S_n] The sum of the terms of a sequence from the first term to the nth term.

Ex: Given the sequence 5, 10, 20, 40, 80, 160,
   a) the third partial sum is given by $S_3 = 5 + 10 + 20 = 35$.
   b) the fifth partial sum is given by $S_5 = 5 + 10 + 20 + 40 + 80 = 155$.
   c) the first partial sum is just the first term, or $S_1 = 5$

In Exercises #3-6, find the indicated partial sum for the given sequence.

3. 10, 8, 6, 4, ..., find S_5 4. $a_n = (n+1)^2 - 3$, find S_3

5. $\frac{1}{3}, \frac{2}{3}, 1, \frac{4}{3}, ...$, find S_7 6. $\begin{aligned} a_1 &= 6 \\ a_n &= 0.5a_{n-1} \end{aligned}$, find S_4

<u>**Sequence of Partial Sums**</u>

Sequence of Partial Sums: A sequence of partial sums is a sequence where the nth term is the partial sum S_n. In this way, the sequence of partial sums is like a "running total so far" for the terms in the original sequence.

Ex: For the sequence 5, 10, 20, 40, 80, 160, the sequence of partial sums is S_1, S_2, S_3, S_4, S_5, S_6, or 5, 15, 35, 75, 155, 315.

---

7. Given the sequence 1, 4, 9, 16, ... , write the first five terms for the sequence of partial sums

8. If $a_n = -7 + \frac{4}{5}(n-1)$, write the first five terms for the sequence of partial sums

9. A basketball tournament is held that includes 64 teams. Each round the teams are paired off and play with the loser being eliminated from the tournament.
   a. The sequence representing the number of games played in each round of the tournament begins 32, 16, ... Complete the sequence and explain why it's a finite sequence.

   b. Turn the sequence into a finite series and find the sum of the series. Explain what this number represents and why it might be useful to the people running the tournament.

   c. Write the sequence of partial sums and explain what it represents in this context.

---

10. Suppose a pattern is formed using blocks as follows. Let n represent the step number, let a_n represent the number of blocks added at Step n, and let S_n represent the total number of blocks at Step n.

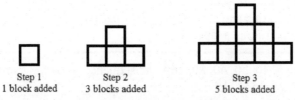

Step 1
1 block added

Step 2
3 blocks added

Step 3
5 blocks added

Then $a_n = 2n - 1$ represents the number of blocks added at Step n.

a. Write the first five terms for a_n and S_n, then explain what the terms of these sequences represent.

b. Write a formula that determines S_n given a step number n.

c. How many blocks will be added at step 9? How many total blocks are in Step 9?

d. Which step number has 225 total blocks?

> ## Finite Arithmetic Series
>
> A *finite arithmetic series* is the sum of the terms of a finite arithmetic sequence.
>
> ---
>
> *Ex:* $-7, -2, 3, 8, 13$ is a finite arithmetic sequence because it has a common difference ($d = 5$) and only finitely many terms. Then $(-7) + (-2) + 3 + 8 + 13$ is the corresponding finite arithmetic series with a sum of 15.

1. a. Is the following series arithmetic? If so, write down the common difference. If not, explain why the series is not arithmetic.
$$1 + 3 + 5 + 7 + 9 + 11 + 13$$

    b. What is the sum of the series?

    c. Starting with the arithmetic series in part (a), suppose we duplicate the series exactly and combine the two series as shown.
$$1 + 3 + 5 + 7 + 9 + 11 + 13 + 1 + 3 + 5 + 7 + 9 + 11 + 13$$
How is the sum of *all* of the terms of the two series combined related to the sum of the original series?

    d. Is your answer to part (c) true even if we change the order of terms? For example, is it still true if we write the following?
$$1 + 3 + 5 + 7 + 9 + 11 + 13 + 13 + 11 + 9 + 7 + 5 + 3 + 1$$

Let's visualize the terms of the arithmetic series as the heights of vertical bars. For example, the terms of the series $1 + 3 + 5 + 7 + 9 + 11 + 13$ are represented below on the left. We have also duplicated the series to the right.

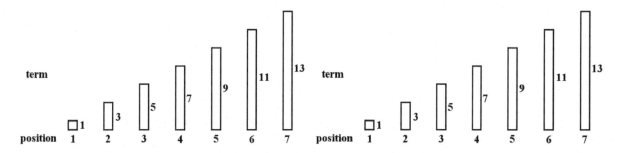

Let's rotate the duplicated terms 180° and then stack them on the terms of the original series.

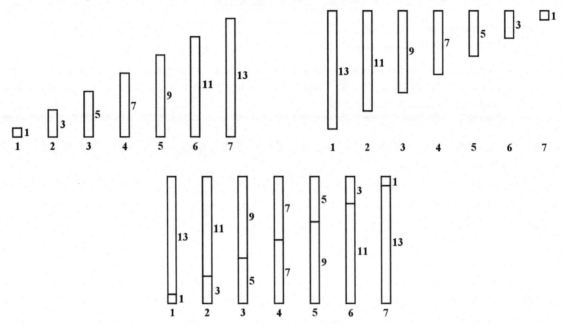

This visualization can help us find the sum of all of the terms in both series together very easily.

2. a. What is the height of each combined bar?   b. How many bars are there?

    c. What is the total sum of all of the terms in both series together and how is this represented in the diagram?

    d. What is the sum of just the original series $1+3+5+7+9+11+13$?

The arithmetic series $1+3+5+7+9+11+13$ is very short and easy to add up on a calculator. However, the reasoning we just explored can really help us when the series get very long. Let's continue to explore this idea.

3. a. The terms in the arithmetic series $2+5+8+11+14+17$ are represented visually below. Imagine duplicating the series and visualizing it as vertical bars rotated 180° and stacked on top of the original diagram. Draw this.

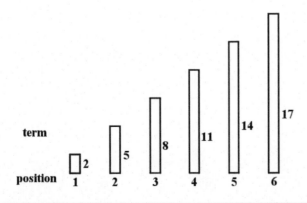

        b. How tall is each bar?

        c. How many bars are there?

        d. What is the total sum of all of the terms in both the original and duplicated series together?

        e. What is the sum of the original arithmetic series $2+5+8+11+14+17$?

The visualization we've used so far is a really nice geometric argument for the technique we're developing. However, let's repeat the reasoning without the same visual component.

4. a. The series $1+3+5+7+9+11+13$ from Exercises #1-2 is shown below. In addition, the series is duplicated and written in reverse order on the next line. Total each column and put your answers in the given boxes.

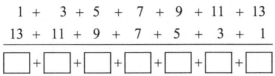

$$1 + \quad 3 + 5 + \quad 7 + \quad 9 + 11 + 13$$
$$13 + 11 + 9 + \quad 7 + 5 + \quad 3 + \quad 1$$
$$\boxed{}+\boxed{}+\boxed{}+\boxed{}+\boxed{}+\boxed{}+\boxed{}$$

  b. What is the sum of every column and how many columns are there?

  c. What is sum of all the terms in both series combined?

  d. What is the sum of the arithmetic series $1+3+5+7+9+11+13$?

5. Use the reasoning we've developed to find the sum of each of the following arithmetic series.
    a. $105+141+177+213+249+285+321$ b. $4+7+10+13+16+19+22+25$

    c. $15+13+11+9+7+5+3+1-1-3-5-7$ d. $-126-117-108-99-90-81-72-63-54$

6. Does the technique we've been developing work with non-arithmetic series? Let's find out. Try to use the same reasoning to find the sum of each of the following series. *Note: You should verify that each series is <u>not</u> arithmetic.*
    a. $1+4+9+16+25+36+49$ b. $2+4+8+16+32+64+128$

7. Why does this technique only work with arithmetic series? Create a convincing mathematical argument.

8. a. To develop a general formula we need to think about how to apply the reasoning to ALL arithmetic series. Suppose that $a_1 + a_2 + a_3 + ... + a_{n-2} + a_{n-1} + a_n$ is an arithmetic series with n total terms. Duplicate the series and write it in reverse order below the original series.

$$a_1 \ + \ a_2 \ + \ a_3 \ + \ ... \ + \ a_{n-2} \ + \ a_{n-1} \ + \ a_n$$

b. What is the sum of each column/pair of terms?

c. What the total combined sum of all of the terms in both series together?

d. What is the sum of the original arithmetic series with n terms?

Sum of a Finite Arithmetic Series

If $a_1 + a_2 + a_3 + ... + a_n$ is a finite arithmetic series, then the sum of the series S_n is...

---

9. Recall the following context from Investigation 3. *Auditoriums are often designed so that there are fewer seats per row for rows closer to the stage. Suppose you are sitting in Row 22 at an auditorium and notice that there are 68 seats in your row. It appears that the row in front of you has 66 seats and the row behind you has 70 seats. Assume this pattern continues throughout the auditorium.*

   a. Go back to Investigation 3 and locate the following information: the number of rows, the number of seats in the first row, and the number of seats in the last row.

   b. How many seats are in the auditorium?

10. One method of exercising with weights is to complete 10 repetitions, rest for 30 seconds, and then perform the same exercise completing 9 repetitions. After a 30 second rest, 8 repetitions are completed. This process continues until the person only lifts the weight once. How many total repetitions are completed during such a routine?

11. Shilah recently decided to start jogging each morning and wants to work up to 3.5 miles per day. Tomorrow she will run 0.5 miles, and she will increase the distance she runs by 0.2 miles each day until she reaches 3.5 miles per day.

   a. How many days will she have been jogging when she reaches 3.5 miles per day?

   b. How many total miles will Shilah have run during this time?

In Exercises #12-17, do the following.
  a) Verify that the series is arithmetic. [*Hint: It might help to write out the first several terms of the series if they are not given.*]
  b) If the series is arithmetic, find its sum without simply adding up all of the terms. [*Hint: In most cases you will need to determine the number of terms in the series.*]

12. The sum of integers from 1 to 100.

13. Find the sum of the integers from 1 to 250.

14. $6 + 13 + 20 + ... + 181$

15. $14 + 10 + 6 + ... - 138$

16. the sum of the even numbers from 1 to 300

17. the sum of the positive four-digit multiples of 20

18. Write a finite arithmetic series with a sum of 80 that has 10 terms and a common difference $d \neq 0$. (*Instead of using "guess and check" try to think about a strategy for solving this problem using the ideas you've learned in this investigation.*)

> ### Finite Geometric Series
>
> A *finite geometric series* is the sum of the terms of a finite geometric sequence.
> ___
> *Ex:* 1, –2, 4, –8 is a finite geometric sequence because it has a common ratio ($r = -2$) and only finitely many terms. Then $1 + (-2) + 4 + (-8)$ or $1 - 2 + 4 - 8$ is the corresponding finite geometric series with a sum of –5.

For Exercises #1-3, consider the finite geometric series $5 + 15 + 45 + 135 + 405 + 1,215$ with a sum $S_6 = 1,820$.

1. a. Suppose each of the terms of the series is multiplied by 2. Write out the terms of the new series, then find its sum and compare this to the sum of the original series.

    b. Suppose each of the terms of the series is multiplied by 4. Write out the terms of the new series, then find its sum and compare this to the sum of the original series.

2. If we multiply each term of the original series by 3, we get the following new series.
    $$3(5) + 3(15) + 3(45) + 3(135) + 3(405) + 3(1,215)$$
    $$15 + 45 + 135 + 405 + 1,215 + 3,645$$

    a. How does the sum of this new series compare to the sum of the original series?

    b. Compare the terms of the new series and the terms of the original series. What do you notice?

    c. Why do so many of the terms from the original series reappear in the new series when the term values are multiplied by 3, but not by 2 or 4?

3. Let S_6 represent the sum of the original series and $3 \cdot S_6$ represent the sum of the new series.

   a. What is the value of $S_6 - (3 \cdot S_6)$?

   b. One of your classmates claims that he can calculate $S_6 - (3 \cdot S_6)$ without knowing the sums of the series. He says that $a_1 - 3a_6$ will have the same value as $S_6 - (3 \cdot S_6)$ and shows the following work. Is he correct?

   $$S_6 = 5 + \cancel{15} + \cancel{45} + \cancel{135} + \cancel{405} + \cancel{1,215}$$

   $$-\left(3 \cdot S_6 = \quad \cancel{15} + \cancel{45} + \cancel{135} + \cancel{405} + \cancel{1,215} + 3,645\right)$$

   $$S_6 - 3 \cdot S_6 = 5 - 3,645$$

   $$S_6 - 3 \cdot S_6 = -3,640$$

   c. Use the equation $S_6 - 3 \cdot S_6 = -3,640$ to solve for S_6.

4. Let's practice the technique from Exercise #3 with the finite geometric series
   $2 + 10 + 50 + 250 + 1,250 + 6,250 + 31,250$ with a sum S_7 and a common ratio $r = 5$. If we multiply the terms of the series by 5 we get the following series with a sum of $5 \cdot S_7$.

   $$2(5) + 10(5) + 50(5) + 250(5) + 1,250(5) + 6,250(5) + 31,250(5)$$
   $$10 + 50 + 250 + 1,250 + 6,250 + 31,250 + 156,250$$

   Complete the steps to find S_7 without actually adding up the terms of the series.

   $$S_7 = 2 + 10 + 50 + 250 + 1,250 + 6,250 + 31,250$$
   $$-\left(5 \cdot S_7 = \quad 10 + 50 + 250 + 1,250 + 6,250 + 31,250 + 156,250\right)$$
   $$S_7 - 5 \cdot S_7 = \rule{3cm}{0.4pt}$$

   $$\rule{1.5cm}{0.4pt} \, S_7 = \rule{3cm}{0.4pt}$$

   $$S_7 = \rule{3cm}{0.4pt}$$

Sum of a Finite Geometric Series

If S_n is the sum of a finite geometric series with a first term a_1, a last term a_n, and a common ratio r, then

We then solve the equation for S_n to find the sum.

This technique allows us to find the sum
- without adding up all of the term values and
- even if we don't know all of the term values or the number of terms.

In Exercises #5-10, find the sum of each finite geometric series.

5. $2 + 4 + 8 + ... + 2,048$

6. $1 + 6 + 36 + ... + 10,077,696$

7. The series begins $2 - 6 + 18 - ...$ and there are 11 terms.

8. $\frac{1}{16} - \frac{1}{4} + 1 - ... - 1024$

9. $4 + \frac{4}{3} + \frac{4}{9} + \frac{4}{27} + \ldots + \frac{4}{2,187}$

10. Rising healthcare costs are a constant issue facing families and businesses. Suppose the cost of insuring an individual is \$8,900 this year and is expected to increase by 7% per year for the next 9 years. How much do we expect it to cost to cover this person over the entire 10-year period?

So far we have written series in a form like $4 + 7 + 10 + 13 + 16$. However, this form has shortcomings. Series with many terms are very common in math, and in these cases we'd need to either write out the entire long series or shorten it (like writing $12 + 17 + 22 + ... + 122$) which can hide important information. Mathematicians needed a way to write the sum of long series in a condensed way that also provides maximum information. Their solution was ***sigma notation***.

The best way to introduce sigma notation is through an example. Consider the series $4 + 7 + 10 + 13 + 16$. This series is made up of the terms of the sequence $4, 7, 10, 13, 16$ described with the explicit formula $a_n = 4 + 3(n-1)$. The first term is $a_1 = 4$. The last term is $a_5 = 16$. We use the capital Greek letter sigma (Σ) to mean *sum of*.

We want to represent the statement "the sum of the terms of the sequence a_n from a_1 to a_5" in some way other than $4 + 7 + 10 + 13 + 16$. We can write this series in sigma notation as $\sum_{n=1}^{5} [4 + 3(n-1)]$.

Sigma Notation

We can write a series with terms a_n in sigma notation as follows.

"to a_5"

"the sum of" $\longrightarrow \sum_{n=1}^{5} a_n \longleftarrow$ "the term values a_n of the sequence"

"from a_1"

If we know the **explicit** formula defining the terms of the series (such as $a_n = 4 + 3(n-1)$), then we can replace a_n with its explicit formula.

$$\sum_{n=1}^{5} [4 + 3(n-1)]$$

Here are two important points about sigma notation.

I) $\sum_{n=1}^{5} a_n = a_1 + a_2 + a_3 + a_4 + a_5$. We can translate sigma notation into the expanded form by substituting $n = 1$, then continuing to substitute values of n until we reach the index value written above sigma (in this case $n = 5$).

II) The number written above sigma is **not** intended to represent *the number of terms*. For example, consider $\sum_{n=5}^{8} a_n$ (4 terms) and $\sum_{n=2}^{3} a_n$ (2 terms).

$$\sum_{n=5}^{8} a_n = a_5 + a_6 + a_7 + a_8 \qquad\qquad \sum_{n=2}^{3} a_n = a_2 + a_3$$

In Exercises #1-4 , write out the terms of the series and find the sum.

1. $\sum_{n=1}^{4} [22 - 2(n-1)]$

    $22, 20, 18, 16$

2. $\sum_{n=1}^{5} 4(3)^{n-1}$

    $4, 12, 36, 108, 324$

3. $\sum_{n=1}^{3} 2^{n+2}$

    $8, 16, 32$

4. $\sum_{n=3}^{7} \frac{n^2}{2}$

    $\frac{9}{2}, \frac{16}{2}, \frac{25}{2}, \frac{36}{2}, \frac{49}{2}$

    $4.5, 8, 12.5, 18, 24.5$

In Exercises #5-8, write each series in sigma notation. (*Hint: You might first need to determine the number of terms in the series.*)

5. $3+15+75+375+1,875+9,375$

$$\sum_{n=1}^{6} 3(5)^{n-1}$$

6. $2+6+10+14+...+90$

$$\sum_{n=0}^{22} 2+4n$$

OR

$$\sum_{n=1}^{23} 2+4(n-1)$$

7. $1+\frac{4}{3}+\frac{5}{3}+2+...+33$

$$\sum_{n=0}^{96} 1+\frac{1}{3}n$$

8. $\dfrac{1}{2^3}+\dfrac{1}{2^4}+\dfrac{1}{2^5}+...+\dfrac{1}{2^{10}}$

$$\sum_{n=3}^{10} \frac{1}{2^n}$$

9. Represent the sum of the first 150 odd positive integers using sigma notation. *Hint: It might help to write out the first several terms of the series first.*

10. Represent the sum of all of the numbers from 1 to 400 that are evenly divisible by 5. *Hint: It might help to write out the first several terms of the series first.*

11. Recall the following context from Investigations 3 and 6. *Auditoriums are often designed so that there are fewer seats per row for rows closer to the stage. Suppose you are sitting in Row 22 at an auditorium and notice that there are 68 seats in your row. It appears that the row in front of you has 66 seats and the row behind you has 70 seats. Assume this pattern continues throughout the auditorium.*

Represent the total number of seats in the auditorium using sigma notation. *(You can refer back to Investigations 3 and 6 to find any additional information you may need.)*

12. Recall the following context from Investigation 5. *A basketball tournament is held that includes 64 teams. Each round the teams are paired off and play with the loser being eliminated from the tournament. The sequence representing the number of games played in each round of the tournament begins 32, 16, ...*

Represent the total number of games played in the tournament using sigma notation. *(You can refer back to Investigation 5 to find any additional information you may need.)*

13. *Recall the following context from Investigation 7. Rising healthcare costs are a constant issue facing families and businesses. Suppose the cost of insuring an individual is $8,900 this year and is expected to increase by 7% per year for the next 9 years.*

Represent the expected cost to cover this person over the entire 10-year period using sigma notation.

In Exercises #14-17, do the following.
    a) State whether the series is arithmetic, geometric, or neither.
    b) If the series is arithmetic or geometric, give the common difference or common ratio.
    c) Write down the first and last terms of the series.

14. $\displaystyle\sum_{n=1}^{14}[3+2n]$ 15. $\displaystyle\sum_{n=1}^{6}7^{n}$ 16. $\displaystyle\sum_{n=1}^{11}\frac{1}{n^{2}}$ 17. $\displaystyle\sum_{n=5}^{9}[4-6(n-1)]$

For Exercises #18-21, find the sum of the series.

18. $\displaystyle\sum_{i=1}^{10}4(3)^{i-1}$ 19. $\displaystyle\sum_{n=1}^{105}[6-4(n-1)]$

20. $\displaystyle\sum_{n=1}^{5}\left[n^{2}-6\right]$ 21. $\displaystyle\sum_{n=1}^{40}3(n-3)$

22. When writing a series using sigma notation we use the explicit formula and not the recursive formula. For example, $12+17+22+...+122$ is written as $\displaystyle\sum_{n=1}^{28}[12+5(n-1)]$ and not as $\displaystyle\sum_{i=1}^{28}[a_{n-1}+5]$. What are the advantages of using the explicit formula for the term values instead of the recursive formula in sigma notation?

1. Explain what it means for a sequence to have a limit.

2. Explain what the sequence of partial sums represents for a series.

3. Explain what each of the following represents.

   a. $\displaystyle\sum_{n=6}^{10}[3n+2]$

   b. $\displaystyle\sum_{n=1}^{\infty}\left[40+n^2\right]$

4. Using a ruler, draw a square on a piece of paper. Make sure that the square is reasonably large – at least 4 inches per side. We'll call the area of this square "one square unit." Now consider the series
$$\tfrac{1}{2}+\tfrac{1}{4}+\tfrac{1}{8}+\tfrac{1}{16}+\ldots$$
   Let's imagine these terms each represent areas (in square units).
   a. Write the first six terms of the sequence of partial sums.

   b. Cut your square in half parallel to one of the sides. You are now holding two rectangles, each with an area of ½ square unit. Set one of these to the side. This represents the value of the first term in your sequence of partial sums.

   Take the second rectangle and cut it in half. You are now holding two squares, each with an area of ¼ square unit. Set one of these aside next to the rectangle with an area of ½ square unit. Repeat this process several more times. How does this activity represent the sequence of partial sums?

   c. Does the sequence of partial sums have a limit? If so, what is the limit?

5. Consider the series $6 + 2 + \frac{2}{3} + \frac{2}{9} + \ldots$. This time, let's imagine the terms represent lengths.

   a. Write the first six terms in the sequence of partial sums.

   b. On the line below, a bolded line segment is drawn measuring 6 units. This represents the first term in the sequence of partial sums. Extend this line segment to generate a segment whose length is the second term in the sequence of partial sums.

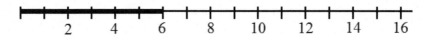

   c. Repeat part (b) to represent the next several terms in the sequence of partial sums.

   d. Does the sequence of partial sums have a limit? If so, what is the limit?

For Exercises #6-11, do the following.

   a) State whether the series is arithmetic or geometric and identify the common difference or common ratio.

   b) Graph the first six terms of the sequence that makes up the series. Does there appear to be a limit? If so, what is it?

   c) Generate and graph the first six terms of the sequence of partial sums for the series.

   d) Does the sequence of partial sums appear to have a limit? If so, what is it?

6. $\displaystyle\sum_{n=1}^{\infty} 2^{n+3}$

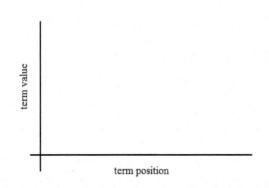

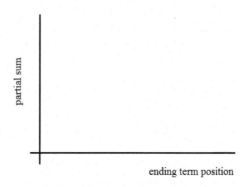

7. $-1 + 2 + 5 + ...$

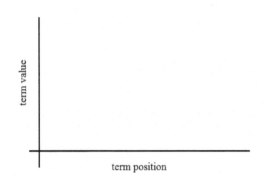

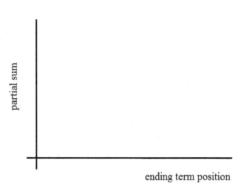

8. $\displaystyle\sum_{n=1}^{\infty} 3\left(-\tfrac{1}{3}\right)^{n-1}$

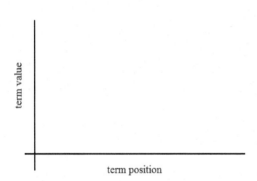

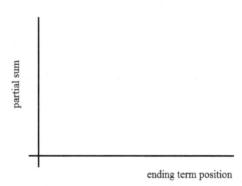

9. $\displaystyle\sum_{n=1}^{\infty} -2n + 6$

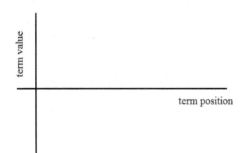

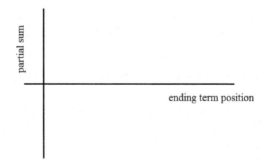

10. $3 + \frac{3}{2} + \frac{3}{4} + \frac{3}{8} + ...$

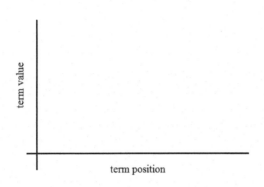 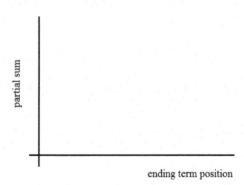

11. $4 - 8 + 16 - 32 + ...$

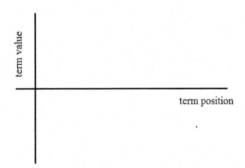 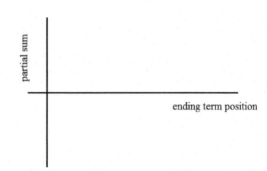

12. What does it mean when the sequence of partial sums for an infinite series has a limit? (In other words, why is part (d) important in Exercises #4-9?)

13. TRUE or FALSE: For an infinite geometric series with a common ratio $r \neq 0$ and $a_1 \neq 0$, the sequence of partial sums always has a limit. Justify your answer.

Convergence and the Sum of an Infinite Series

Sum of an Infinite Geometric Series

If S is the sum of an infinite geometric series with a first term a_1 and a common ratio r, then

We solve the equation for S to find the sum.

Remember:
- The sum of an infinite series is the limit of the sequence of partial sums.
- An infinite geometric series will only have a sum if $|r| < 1$.

In Exercises #1-3, find the sum if possible.

1. $\displaystyle\sum_{n=1}^{\infty} 24\left(\tfrac{2}{3}\right)^{n-1}$

2. $\displaystyle\sum_{n=1}^{\infty} (1.2)^{n}$

3. $\displaystyle\sum_{n=1}^{\infty} 7(-0.4)^{n-1}$

4. A pendulum is made up of a string or solid arm with a weight attached to the end. Because of outside factors such as air resistance and gravity, each swing of a pendulum is a little shorter than the previous one. Suppose the length of the pendulum's swing follows a geometric sequence with the first swing being 100 cm long and the second swing length being 99% of the previous swing length.

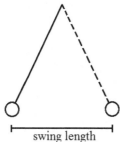

swing length

a. What is the common ratio for the sequence of pendulum swing lengths? Write out the first several terms of the sequence of swing lengths.

b. Suppose an infinite series is formed from this sequence. Find the sum of the series (if possible).

c. Does the sum of the series have a real-world meaning? If so, describe the meaning. If not, explain your thinking.

For Exercises #5-8, do the following.
  a) Write the series in sigma notation.
  b) Find the sum of the series (if possible).

5. $120 + 30 + \frac{30}{4} + \frac{30}{16} + ...$

6. $-10 + 5 - 2.5 + 1.25 - ...$

7. $1.25 - 2.5 + 5 - 10 + ...$

8. $1 + x + x^2 + x^3 + ...$ where $-1 < x < 1$

In Exercises #9-11, create a series that meets the stated requirements. Write the series using sigma notation. (*Note: Do not use any of the series we have provided in this investigation.*)

9. an infinite geometric series with a negative common ratio that has a sum

10. an infinite geometric series that does not have a sum

11. an infinite geometric series with a sum between 0 and 1

12. A doctor prescribes a pain reliever to a patient with severe back pain. The doctor tells the patient to take a 500 mg does every 8 hours and that she will reevaluate the patient in the future. After some research about the medication, the patient learns that his body likely eliminates 90% of the drug in his system over the course of 8 hours.

a. How much of the drug is in the patient's system right after he takes the second dose? Show the expression that calculates this value.

b. How much of the drug is in the patient's system right after he takes the third dose? Show the expression that calculates this value.

c. How much of the drug is in the patient's system right after he takes the fourth dose? Show the expression that calculates this value.

d. Explain how the amount of drug in the patient's system after the nth dose forms a geometric series.

e. Suppose the patient stays on the medication indefinitely. What eventually happens to the amount of drug in his system when he takes a new dose?

13. Suppose the doctor prescribes an ibuprofen regimen to another patient to alleviate problems with swollen joints. The doctor tells the patient to take 600 mg of ibuprofen every 12 hours, and during each 12-hour period the patient's body eliminates 85% of the drug in the patient's system.
    a. Find the amount of drug in the patient's system right after taking the 7th dose.

    b. Suppose the patient stays on the medication indefinitely. What eventually happens to the amount of drug in his system?

INVESTIGATION 1: SEQUENCES AND THE LIMITS OF SEQUENCES

For Exercises #1-6, do the following.
  a) Describe the pattern.
  b) Complete the table showing the relationship between the term positions and the term values.

input: term position	output: term value
1	
2	
3	
4	
5	
6	

  c) Graph the sequence (use at least six points).
  d) Does the sequence appear to have a limit? If so, what is it?

1. $40, 52, 64, ...$

2. $15, 5, \frac{5}{3}, ...$

3. $1, \frac{2}{3}, \frac{4}{9}, ...$

4. $-6, -7.8, -9.6, ...$

5. $0, 1, 4, 9, ...$

6. $6, 9, 10.5, 11.25, ...$

7. A sequence is formed by choosing any real number x to be the first term, and then generating the sequence by taking each term value and dividing by 2 to get the next term value.
  a. Choose a few possible values of x and explore the kinds of sequences produced by following this pattern.
  b. Do all sequences formed in this manner have a limit? If so, what is the limit? Why does this happen?

8. A sequence is formed by choosing any real number x to be the first term, then generating the sequence by taking each term value and multiplying by 3.5 to get the subsequent term value.
  a. Choose a few possible values of x and explore the kinds of sequences produced by following this pattern.
  b. Do all sequences formed in this manner have a limit? If so, what is the limit? Why does this happen?

9. A pyramid is formed from blocks according to the given pattern. Write the next three terms in the sequence showing the total number of blocks in the pyramid at each stage.

 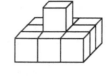 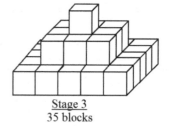

Stage 1
1 block

Stage 2
10 blocks

Stage 3
35 blocks

10. A sequence is formed by determining the number of diagonals that can be drawn inside of regular polygons as shown below. (A diagonal is a line segment drawn inside of the figure from one vertex to another.) Determine the pattern formed by this sequence and write the next three terms of the sequence.

Stage 1

Stage 2

Stage 3

Stage 4

0 diagonals

2 diagonals

5 diagonals

??? diagonals

11. An infinite sequence begins 40, 60, ...
    a. Determine a possible pattern for this sequence such that the sequence *will not have a limit*. Write the next three terms using this pattern.
    b. Determine another possible pattern for this sequence such that sequence *will not have a limit*. Write the next three terms using this pattern.
    c. Determine another possible pattern for this sequence such that sequence <u>*will*</u> *have a limit*. Write the next three terms using this pattern.

12. An infinite sequence begins $-8, -3$, ...
    a. Determine a possible pattern for this sequence so that the sequence *will not have a limit*. Write the next three terms using this pattern.
    b. Determine another possible pattern for this sequence so that sequence *will not have a limit*. Write the next three terms using this pattern.
    c. Determine another possible pattern for this sequence so that sequence <u>*will*</u> *have a limit*. Write the next three terms using this pattern.

INVESTIGATION 2: FORMULAS FOR SEQUENCES

In Exercises #13-16, find the indicated term value for the given sequence.

13. $1, -3, 9, \ldots$;   find a_5

14. $104, 107, 110, \ldots$;   find a_6

15. $\frac{1}{2}, \frac{1}{3}, \frac{1}{4}, \ldots$;   find a_{10}

16. $1, -5, 25, -125, 625, \ldots$;   find a_5

17. If a_n represents the value of the nth term in some sequence, how could you represent each of the following?
    a. the value of the term three term positions before the nth term
    b. the value of the term two term positions after the nth term
    c. the term position one term prior to the nth term

18. If a_n represents the value of the nth term in some sequence, how could you represent each of the following?
    a. the term position four terms after the nth term
    b. the value of the term five term positions before the nth term
    c. the value of the term three term positions after the nth term

In Exercises #19-24, use the given formula to write the first four terms of each sequence.

19. $a_1 = -2$, $a_n = a_{n-1} + 3$

20. $a_1 = 360$, $a_n = \dfrac{a_{n-1}}{3}$

21. $a_1 = 10$, $a_n = -2a_{n-1} + 4$

22. $a_1 = 4$, $a_n = \dfrac{128}{a_{n-1}}$

23. $a_1 = -1$, $a_2 = 4$
$a_n = a_{n-1} + a_{n-2}$

24. $a_1 = 1$, $a_2 = 2$, $a_3 = 4$
$a_n = 2a_{n-3} + 3a_{n-2}$

In Exercises #25-28, write a recursive formula to define each sequence.

25. $25, 30, 35, \ldots$

26. $1, -7, -15, \ldots$

27. $3, 18, 108, \ldots$

28. $1, -\frac{1}{8}, \frac{1}{64}, \ldots$

In Exercises #29-30, write the first four terms of the sequence.

29. $a_n = \dfrac{5(n-1)}{4n}$

30. $a_n = 7(-3)^n$

In Exercises #31-32, find the value of the 15ᵗʰ term in the sequence.

31. $a_n = -4 - 15(n-1)$

32. $a_n = \dfrac{14}{(n+3)^2}$

In Exercises #33-34, write the explicit formula defining the sequence and then find the number of terms in the given sequence.

33. $1, 4, 9, 16, 25, ..., 225$

34. $10, 13, 16, 19, ..., 178$

In Exercises #35-38 you are given the explicit formula for a sequence. Write the inverses for each relationship. (*That is, write the formulas the represent the term's position based on its value.*)

35. $a_n = 3n + 6$ 36. $a_n = -2(n-1) + 8$ 37. $a_n = n^2 + 7$ 38. $a_n = \frac{6n-12}{5}$

39. After a car is purchased its value generally decreases over time. Suppose a car's original purchase price is \$24,679 and that is the value of the car in any given year is always 0.85 times as large as the value of the car the year before.
   a. Write a sequence showing the car's value at the end of each of the first five years since its original purchase.
   b. Write a recursive formula for the sequence in part (a).
   c. Write an explicit formula for the sequence in part (b).

40. Suppose we have 57 micrograms of a particular bacteria population in a petri dish. We place this petri dish in an incubator and the bacteria grows in such a way that its mass triples each day.
   a. Write a sequence showing the bacteria's mass at the end of each of the first five days since it was placed in the incubator.
   b. Write a recursive formula for the sequence in part (a).
   c. Write an explicit formula for the sequence in part (b).

INVESTIGATION 3: ARITHMETIC SEQUENCES

41. Your friends learned about sequences in another class and don't understand how or why the explicit formula for arithmetic sequences $(a_n = a_1 + d(n-1)$ or $a_n = d(n-1) + a_1)$ works to find any term value (they just memorized the formula for a test). Explain to your friends what each of the following represents in the formula. *Hint: Think about linear functions.*
   a. n b. d c. a_n d. $n-1$ e. $d(n-1)$ f. $a_1 + d(n-1)$ or $d(n-1) + a_1$

42. The explicit formula for the term values of a certain sequence is $a_n = 4(n-1) + 19$. The calculations to determine the value of the 33ʳᵈ term are given. Answer the questions that go with these calculations.

$a_n = 4(n-1) + 19$

$a_{33} = 4(33-1) + 19$ • Step 1

$a_{33} = 4(32) + 19$ • Step 2

$a_{33} = 128 + 19$ • Step 3

$a_{33} = 147$ • Step 4

   a. What do "4" and "19" represent?
   b. In Step 1, what does the expression 33 – 1 represent?
   c. In Step 2, what does the expression 4(32) represent?

For Exercises #43-50, do the following.
   a) Write the recursive formula for the arithmetic sequence.
   b) Write the explicit formula for the arithmetic sequence.
   c) Determine the number of terms in the arithmetic sequence.

43. $13, 15, 17, ..., 251$

44. $14, 13.7, 13.4, ..., -10.6$

45. $62, 63.1, 64.2, ..., 124.7$

46. $-30, -26.5, -23, ..., 488$

47. $-33, -41, -49, ..., -825$

48. $1077, 1077.1, 1077.2, ..., 1085.4$

49. All of the 3-digit multiples of 7

50. All of the 5-digit multiples of 25

51. Interior angles of convex polygon are formed by two adjacent sides of the polygon opening into the interior of the figure. The sum of the measures of all of the interior angles of a polygon follow the pattern shown below.

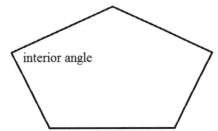

number of sides n	sum of the measures of the interior angles a_n
3	180°
4	360°
5	540°
6	720°

   a. We can think of the sequence beginning with a_3 instead of a_1. Write the recursive formula for the sequence.
   b. Write the explicit formula for the sequence.
   c. Regular polygons are polygons where all of the interior angles are congruent and all of the sides are the same length. What is the measure of *each* interior angle in a regular polygon with 18 sides?

52. Auditoriums are often designed so that there are fewer seats per row for rows closer to the stage. Suppose you are sitting in Row 18 at an auditorium and notice that there are 73 seats in your row. It appears that the row in front of you has 70 seats and the row behind you has 76 seats. Suppose this pattern continues throughout the auditorium.
   a. How many seats are in the first row?
   b. How many seats are in the 30th row?
   c. Write an explicit formula for the sequence where a_n is the number of seats in the nth row.

In Exercises #53-54, solve the given explicit formula for n and explain what questions this new version of the formula helps you to answer.

53. $a_n = 6(n-1) - 27$

54. $a_n = 6(n-1) - 27$

[*Systems of Equations*] In Exercises #55-56 you are given the explicit formulas for the term values of pairs of sequences. For each pair, is there a term in one sequence that shares the same value and term position as a term in the other sequence? If so, give their value and position.

55. $a_n = -3(n-1) + 30$ and $b_n = 2(n-1) - 25$

56. $a_n = 4(n-1) + 25$ and $b_n = 6(n-1) - 78$

57. In this module we've seen that it's possible for an infinite sequence to have a limit. However, infinite arithmetic sequences with $d \neq 0$ never have a limit. Explain why this is true.

58. An arithmetic sequence has the following two terms: $a_{13} = 58$ and $a_{28} = 140.5$. Write the explicit formula defining the value of the nth term a_n.

59. An arithmetic sequence has the following two terms: $a_6 = 13.5$ and $a_{24} = 51.3$. Write the explicit formula defining the value of the nth term a_n.

60. A finite arithmetic sequence has the following two terms: $a_9 = 52.9$ and $a_{27} = 121.3$. The value of the final term is 934.5. How many terms are in the sequence?

61. A finite arithmetic sequence has the following two terms: $a_5 = 71.6$ and $a_{23} = 33.8$. The value of the final term is -167.8. How many terms are in the sequence?

INVESTIGATION 4: GEOMETRIC SEQUENCES

In Exercises #62-63 you are given diagrams intended to help a person visualize how the term values of a geometric sequence compare to one another. For each diagram, estimate the value of the common ratio and the values of a_2, a_3, a_4, and a_5.

62.

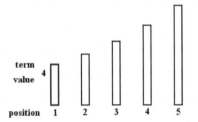

63.

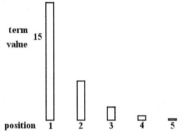

64. Draw a diagram for an arithmetic sequence similar to the diagrams in Exercise #62-63. Compare the diagrams for arithmetic sequences to the diagrams for geometric sequences.

For Exercises #65-70, do the following.
   a) Write the recursive formula for the geometric sequence.
   b) Write the explicit formula for the geometric sequence.
   c) Determine if the sequence has a limit. If so, state the limit.

65. $3, 6, 12, 24, \ldots$

66. $38, 19, 9.5, 4.75, \ldots$

67. $-15, 60, -240, 960, \ldots$

68. $88, -11, \frac{11}{8}, \ldots$

69. $90, 30, 10, \frac{10}{3}, \ldots$

70. $12, 18, 27, 40.5, \ldots$

71. Inflation describes the tendency of prices for goods to rise over time. Historically, prices of goods have risen about 3% per year. Suppose a loaf of bread costs $2.99 in January 2012 and that its cost will rise by 3% per year over the next several years.
   a. Write a recursive formula for the cost of a loaf of bread a_n in dollars n years since January 2012.
   b. Write an explicit formula for the cost of a loaf of bread a_n in dollars n years since January 2012.
   c. Assuming this pattern continues, what do you expect to be the cost of a loaf of bread in 2025?

72. In 2010, a university's budgeting committee decided that they would set the annual tuition rate at $14,000 per year and that they would raise this rate by 2.5% each year following the 2010-2011 school year.
   a. Write a recursive formula for the tuition rate a_n in dollars n years since the 2010-2011 school year.
   b. Write an explicit formula for the tuition rate a_n in dollars n years since the 2010-2011 school year.
   c. Assuming this pattern continues, what do you expect to be the tuition rate for the 2018-2019 school year?

For Exercises #73-76, do the following.
    a) Fill in the blanks to make the sequence arithmetic.
    b) Fill in the blanks to make the sequence geometric.
73. 12, 36, _____   74. _____, 1, −4, ...   75. 14, _____, 224, ...   76. 20, _____, 45, ...

77. A geometric sequence has the following two terms: $a_3 = 72$ and $a_7 = 93,312$. Write the explicit formula defining the value of the nth term a_n.

78. A geometric sequence has the following two terms: $a_5 = 2048$ and $a_8 = 131,072$. Write the explicit formula defining the value of the nth term a_n.

79. A finite geometric sequence has the following two terms: $a_7 = 288$ and $a_{12} = 9,216$. The last term in the sequence is $37,748,736$. How many terms are in the sequence?

80. A finite geometric sequence has the following two terms: $a_4 = 250$ and $a_8 = 156,250$. The last term in the sequence is 2,441,406,250. How many terms are in the sequence?

INVESTIGATION 5: INTRODUCTION TO SERIES AND PARTIAL SUMS

Suppose a series is formed using the terms from the sequences defined in Exercises #81-84 below. For each, find the given partial sum.
81. 103, 206, 309, ..., find S_4 82. 9, 16, 25, 36, ..., find S_6

83. $a_n = \dfrac{(8-n)^2}{n^3}$, find S_1 84. $a_1 = 64$, $a_n = \frac{1}{4}a_{n-1}$, find S_5

In Exercises #85-88, suppose a series is formed using the terms from the given sequence. Write the first five terms for the sequence of partial sums.

85. $a_n = 20 - 6n$ 86. 3, $\frac{9}{2}$, $\frac{27}{4}$, ...   87. $a_1 = 2.5$, $a_n = 6 \cdot a_{n-1} - 2$ 88. $\begin{array}{l}a_1 = 1,\ a_2 = 1\\ a_n = a_{n-2} + a_{n-1}\end{array}$

89. The table below shows the terms of a sequence where a_n is the power bill (in dollars) for a single family home during the nth month of 2011.

n	a_n	S_n	n	a_n	S_n
1	53.15		7	212.63	
2	52.90		8	175.10	
3	57.44		9	118.02	
4	73.80		10	68.98	
5	98.17		11	56.65	
6	152.82		12	53.70	

    a. Fill in the column showing the partial sums S_n.
    b. Explain what the partial sums are keeping track of in this context.

90. The table below shows the terms of a sequence where a_n is the number of DWI (Driving While Intoxicated) tickets given out in Travis County for the nth month of 2012.

n	a_n	S_n
1	62	
2	48	
3	55	
4	41	
5	54	
6	62	

n	a_n	S_n
7	67	
8	70	
9	63	
10	59	
11	56	
12	48	

   a. Fill in the column showing the partial sums S_n.

   b. Explain what the partial sums are keeping track of in this context.

In Exercises #91-93 below you are given information about the terms of arithmetic series and the sequence of partial sums for each series. Complete the tables.

91.

n	a_n	S_n
1		
2		
3	11	
4	15	
5	19	
6		
7		

92.

n	a_n	S_n
1		
2		
3		
4		
5		5
6		0
7		−7

93.

n	a_n	S_n
1		
4		
5		
10	23.5	
13	31	
16		
18		

In Exercises #94-96 below you are given information about the terms of geometric series and the sequence of partial sums for each series. Complete the tables.

94.

n	a_n	S_n
1		
2		
3	54	
4	162	
5		

95.

n	a_n	S_n
1		−20
2		10
3		
4		
5		

96.

n	a_n	S_n
1		
4		
5		
10	2,048	
13	16,384	

INVESTIGATION 6: FINITE ARITHMETIC SERIES

97. a. Draw vertical bars with heights that represent the terms of the series $2+4+6+8+10+12+14$. [*See Exercises #1 and #2 in Investigation 6.*]

   b. Duplicate the diagram and combine it with your diagram in part (a) to demonstrate the total combined sum of the two series. [*See Exercises #1 and #2 in Investigation 6.*]

   c. What is the height of each combined bar and how many bars are there?

   d. What is the total combined sum of the two series?

   e. What is the sum of the original series?

98. Repeat Exercise #97 for the series $1+4+7+10+13+16+19+22$.

99. The series $-4+(-1)+2+5+8+11+14+17$ is shown below. In addition, the series is duplicated and written in reverse order on the next line. Total each column and explain how this can help you determine the sum of the original series.

$$(-4)+ (-1) + \ 2 \ + \ 5 \ + \ 8 \ + 11 + \ 14 + \ 17$$
$$\ 17 \ + \ 14 \ + 11 \ + \ 8 \ + \ 5 \ + \ 2 +(-1) \ +(-4)$$

$$\boxed{}+\boxed{}+\boxed{}+\boxed{}+\boxed{}+\boxed{}+\boxed{}+\boxed{}$$

100. Repeat Exercise #99 for the series $-10+(-6)+(-2)+2+6+10+14+18+22$.

For Exercises #101-102, find the sum of the series.
101. $121+128+135+...+233$ (17 terms)

102. $83+81+79+...-19$ (52 terms)

For Exercises #103-106, do the following for each finite arithmetic series.
   a. Find the number of terms in the series.
   b. Find the sum of the series.
103. $17+21+25+...+89$

104. $223+214+205+...+(-56)$

105. $6+6.3+6.6+...+25.2$

106. $-10+(-8.2)+(-6.4)+...+42.2$

For Exercises #107-108, do the following for each finite arithmetic series.
   a. Give the first three terms of the series and the last term of the series.
   b. Find the number of terms in the series.
   c. Find the sum of the series.
107. the sum of all 4-digit numbers

108. the sum of the 3-digit multiples of 2

109. Auditoriums are often designed so that there are fewer seats per row for rows closer to the stage. Suppose you are sitting in Row 23 at an auditorium and notice that there are 65 seats in your row. It appears that the row in front of you has 63 seats and the row behind you has 67 seats. Suppose this pattern continues throughout the auditorium. You count a total of 42 rows in the auditorium. How many seats does the auditorium contain?

110. A child builds a pyramid out of multiple decks of playing cards. As he starts from the bottom-up, the layers are numbered starting at the base. Suppose he creates the sixth layer by using 12 cards. His friend notes that the fifth layer is made out of 16 cards, and the boy is planning to use 8 cards to create the seventh layer. Suppose this pattern continues throughout the entire pyramid. The boy who is building says that his pyramid will have a total of eight layers. How many cards does it take to build the pyramid?

111. A finite arithmetic series has 73 terms with $a_7=21.4$ and $a_{13}=32.8$. What's the sum of the series?

112. A finite arithmetic series has 90 terms with $a_5=182$ and $a_{86}=291.6$. What's the sum of the series?

113. A finite arithmetic series has a sum of 81,674 with a first term of 912 and a final term of -70. How many terms are in the series?

114. A finite arithmetic series has a sum of 104,030 with a first term of -12 and a final term of 1042. How many terms are in the series?

115. Write a finite arithmetic series with a sum of 200 that has 8 terms with a common difference $d \neq 0$. (*Instead of using "guess and check" try to think about a strategy for solving this problem using the ideas you've learned in this investigation.*)

116. Write a finite arithmetic series with a sum of 540 that has 12 terms with a common difference $d \neq 0$. (*Instead of using "guess and check" try to think about a strategy for solving this problem using the ideas you've learned in this investigation.*)

INVESTIGATION 7: FINITE GEOMETRIC SERIES

117. The finite geometric series $10 + 40 + 160 + \ldots + 163,840$ has 8 terms and a sum S_8. Show the process of multiplying the sum by the common ratio and finding the difference of S_8 and $r \cdot S_8$ to find the sum. (In other words, justify why $S_n - r \cdot S_n = a_1 - r \cdot a_n$ is true for finite geometric series.)

118. The finite geometric series $3 + 9 + 27 + \ldots + 2,187$ has 7 terms and a sum S_7. Show the process of multiplying the sum by the common ratio and finding the difference of S_7 and $r \cdot S_7$ to find the sum. (In other words, justify why $S_n - r \cdot S_n = a_1 - r \cdot a_n$ is true for finite geometric series.)

In Exercises #119-124, find the sum of each finite geometric series.
119. $5 + 25 + 125 + \ldots + 390,625$
120. $-3 + 6 - 12 + \ldots - 786,432$
121. $4 + 12 + 36 + \ldots + 78,732$
122. $-128 + 64 - 32 + \ldots + 1$
123. $1 + \frac{1}{10} + \frac{1}{100} + \ldots + \frac{1}{100,000,000}$
124. $36 + 12 + 4 + \ldots + \frac{4}{2,187}$

125. The local power company plans to raise rates to cover the increased costs of producing power. In a recent newsletter, they notified their customers to expect power costs to increase by 4.2% per year for the next 10 years (including this year). (The power costs one year are 1.042 times as large as the costs in the previous year.)
    a. If you paid a total of $1031.60 for power last year, what do you expect to pay for power over the next 10 years (assuming you continue to live in your current residence and that your power usage does not change)?
    b. How does this compare to the total price if power costs remained the same over the next 10 years?

126. A bouncing ball reaches heights of 16 cm, 12.8 cm, and 10.24 cm on three consecutive bounces.
    a.   If the ball was dropped from a height of 25 cm, how many times has it bounced when it reaches a height of 16 cm? 10.24 cm?
    b.   Write the first five terms of the sequence representing the bounce height after n bounces.
    c.   How much total distance *in the downward direction only* has the ball traveled after five bounces? (*Be careful!*)
    d.   How much total distance *in the upward direction only* has the ball traveled after five bounces (and reaching the top of its fifth bounce)?
    e.   What is the *total distance* traveled by the ball when it reaches the top of its fifth bounce?

127. Suppose you had the option of receiving $100 today, $200 tomorrow, $300 the next day, and so on for a month, or choosing to get $0.01 today, $0.02 tomorrow, $0.04 the next day, $0.08 the day after that, and so. Which would you choose? How much more money do you get with your chosen option? (*Assume "one month" means "30 days"*).

INVESTIGATION 8: SIGMA NOTATION FOR SERIES

In Exercises #128-131, write out the terms of each series.

128. $\displaystyle\sum_{n=1}^{7} 5(2)^{n+1}$

129. $\displaystyle\sum_{n=1}^{5} \frac{n^2 - 3}{10}$

130. $\displaystyle\sum_{n=4}^{8} \frac{4^n + n}{3n}$

131. $\displaystyle\sum_{n=2}^{4} \left(2(5)^n + 3(4)^{n-2}\right)$

In Exercises #132-135, write each series using sigma notation. (*Hint: You might first need to determine the number of terms in the series.*)

132. $10 + 16 + 22 + 28 + 34 + 40 + 46 + 52$

133. $3 - 12 + 48 - 192 + 768 - 3072$

134. $-6 + (-4) + (-2) + ... + 142$

135. $5 + 20 + 80 + ... + 5,242,880$

136. Represent the sum of the numbers from 1 to 300 that are evenly divisible by 3 using sigma notation. *Hint: It might help to write out the first several terms of the series first.*

137. Represent the sum of the numbers from 1 to 1000 that are evenly divisible by 25 using sigma

INVESTIGATION 9: INFINITE SERIES

For Exercises #138-141 do the following.
   a) State whether the series is arithmetic or geometric and identify the common difference or common ratio.
   b) Graph the first six terms of the sequence that makes up the series. Does there appear to be a limit? If so, what is it?
   c) Generate and graph the first six terms of the sequence of partial sums for the series.
   d) Does the sequence of partial sums appear to have a limit? If so, what is it?

138. $\displaystyle\sum_{n=1}^{\infty} 7\left(\frac{4}{3}\right)^{n-1}$

139. $\displaystyle\sum_{n=1}^{\infty} -0.3n + 7$

140. $-1 + 0.2 + (-0.04) + 0.008 + (-0.0016) + 0.00032$

141. $\frac{16}{4}, 5, \frac{25}{4}, \frac{125}{16}, \frac{625}{64}, \frac{3125}{256}$

INVESTIGATION 10: INFINITE GEOMETRIC SERIES

In Exercises #142-144, find the sum if possible.

142. $\displaystyle\sum_{n=1}^{\infty} 28\left(-\frac{5}{4}\right)^{n-1}$

143. $\displaystyle\sum_{n=1}^{\infty} \left(\frac{5}{8}\right)^{n}$

144. $\displaystyle\sum_{n=1}^{\infty} -5(0.95)^{n-1}$

For Exercises #145-148 do the following.
   a) Write the series in sigma notation.
   b) Find the sum of the series (if possible).

145. $-12 + (-6) + (-3) + (-1.5) + ...$

146. $2 - 2.5 + 3.125 - \frac{125}{32} + ...$

147. $2.1 - 2.1x + 2.1x^2 - 2.1x^3 + ...$; where $1 < x < 1.2$

148. $4.5 + 4.5a + 4.5a^2 + ...$; where $|a| < 1$

In Exercises #149-154, create a series that meets the stated requirements. Write the series using sigma notation. (*Note: Do not use any of the series we have provided in the investigations.*)

149. an infinite geometric series with a positive common ratio that does not have a sum

150. an infinite geometric series that has a negative common ratio and does not have a sum

151. an infinite geometric series with a positive common ratio that does have a sum

152. an infinite geometric series with a negative common ratio that does have a sum

153. an infinite geometric series with a sum between 0 and 1

154. an infinite geometric series with a sum between 0 and -1